ERRATA

Page	For	Read
18, line 25	mode.	mean.
20, line 7	x	$x,$
55, line 31	and the	and so the
55, line 32	was one in 999,999, so	or
112, line 35	$X =$	$\bar{X} =$
119, line 16	x	$\bar{x}$
120, line 25	$(x-\mu)$	$(\bar{x}-\mu)$
156, line 14	<+3.182 <+3.182,	<+3.182, or
159, line 17	other is y;	other is $\bar{y}$;
160, line 37	as an actual	as a long run
174, line 12	HERB	HERS
174, line 14	indidence	incidence
184, line 3	a/a+b	$a/(a+b)$
184, line 3	c/c+d	$c/(c+d)$
line 9	$w = a + b \times h,$	$w = \alpha + \beta x \ h$
, line 16	Figure 9.3	Figure 9.5

STATISTICS WITH COMMON SENSE

David Kault

Greenwood Press
Westport, Connecticut • London

Library of Congress Cataloging-in-Publication Data

Kault, David.
    Statistics with common sense / David Kault.
      p.   cm.
    Includes bibliographical references and index.
    ISBN 0–313–32209–0 (alk. paper)
    1. Statistics. I. Title.
  QA276.12.K38   2003
  519.5—dc21   2002075322

British Library Cataloguing in Publication Data is available.

Library of Congress Catalog Card Number: 2002075322
ISBN: 0–313–32209–0

First published in 2003

Greenwood Press, 88 Post Road West, Westport, CT 06881
An imprint of Greenwood Publishing Group, Inc.
www.greenwood.com

Printed in the United States of America

The paper used in this book complies with the
Permanent Paper Standard issued by the National
Information Standards Organization (Z39.48–1984).

10 9 8 7 6 5 4 3 2 1

Contents

Preface

Statistics is primarily a way of making decisions in the face of variability and uncertainty. Often some new treatment is first tried on a few individuals and there seems to be some improvement. We want to decide whether we should believe the improvement is "for real" or just the result of chance variation. The treatment may be some actual medical treatment, or it may be the application of a new fertilizer to a crop or an assessment of the effect of particular social circumstances on social outcomes. In many professional areas people want to answer the same basic question: "Does this make a real difference?" In the modern world this question is answered by statistics.

Statistics is therefore part of the training course for people in a wide range of professions. Sadly, though, statistics remains a bit of a mystery to most students and even to some of their statistics teachers. Formulas and rules are learned that lead to an answer to the question, "Does this make a genuine difference?" in various situations. However, when people actually come to apply statistics in real life they are generally uneasy. They may be uneasy not only because they have forgotten which formula to apply in which situation or which button to press on the computer, but also because the formula or the computer is using criteria that they never properly understood to make important decisions that sometimes don't accord with common sense. People in this situation are right to be uneasy. Statistics applied correctly but without full understanding can lead to the most inappropriate, even bizarre decisions. Common sense without any assistance from statistical analysis will often lead to more sensible decisions. Nevertheless, statistics has conquered the world of modern decision making. Few people notice that many statisticians don't believe in statistics as it is currently practiced. Statistics can of course be used wisely, but this depends on the user properly understanding the meaning of the answers from the formula or the computer and understanding how to combine these answers with common sense.

This book is primarily aimed at people who learned statistics at some stage, never properly understood it, and now need to use it wisely in everyday professional life. However, the book should be equally suitable as an introductory text for students learning statistics for the first time. There is a large number of introductory statistics texts. This text stands out in three ways:

- It emphasizes understanding, not formulas.
- It emphasizes the incorporation of common sense into decision making.
- It gives the full mathematical derivation of some statistical tests to enhance understanding.

The last point requires an immediate qualification to prevent the large number of people with mathematics phobia from shutting the book for good at this point. No mathematical background beyond grade 10 is assumed, and the mathematics often consists of simply explaining one logical idea. Because formulas that can't be fully understood by someone with grade 10 mathematics are omitted, there is less mathematics than in most statistics texts.

The aim is to show the limited connection between wise decision making and statistics as it is conventionally practiced, and to show how this situation can be rectified by combining statistics with common sense.

Acknowledgments

I thank my former statistics teacher, John Hunter, for his teaching, his inspiration, and his suggestions for this book. I also thank my son Sam for his proofreading. I am grateful for the support given to me by James Cook University of North Queensland in writing this book and the accompanying statistical computer program.

Glossaries

MATHEMATICAL SYMBOLS

$<$ less than
$\leq$ less than or equal to
$>$ greater than
$\geq$ greater than or equal to
$\neq$ does not equal
$\approx$ approximately equals
$\sim$ is
$\cup$ or (meaning one or the other or both)
$\cap$ and
$|$ given
$n!$ n factorial, meaning $n \times (n-1) \times (n-2) \times \ldots \times 3 \times 2 \times 1$; for example,

$$4! = 4 \times 3 \times 2 \times 1 = 24$$

$^{n}C_{k}$ Number of ways that from n objects k objects can be chosen (from n Choose k)

$$^{n}C_{k} = \frac{n!}{k!(n-k)!} \; ;$$

for example,

$$^{5}C_{3} = \frac{5 \times 4 \times 3 \times 2 \times 1}{(3 \times 2 \times 1) \times (2 \times 1)} = 10$$

(see Chapter 3).

COMMON ENGLISH EXPRESSIONS USED IN THE
TEXT TO IMPROVE READABILITY

The expressions here on the left-hand side are not normally intended to be used in an absolutely precise way. However, in certain contexts in this book they are used in place of precise quantitative expressions to improve readability. The precise meanings that I attach to these expressions are given on the right.

"hardly ever"	with probability ≤ 0.05
"nearly always"	with probability ≥ 0.95
"quite often" or "commonly"	with probability > 0.05

Statistical Computer Program

Many people frequently come across questionable decisions made on the basis of statistical evidence. This book will help them to make their own informed judgment about evidence based on statistical analyses. Only some people will need to undertake statistical analyses themselves. On the other hand, it is just a small step from understanding statistical evidence to being able to undertake statistical analyses in many situations. It is a small step because in most cases the actual calculations are performed by a computer. The only additional skill to be learned in order for readers to perform statistical analyses for themselves is to learn which button on the computer to press. Doing helps learning, so this book includes questions, some of which are intended to be answered with the assistance of a statistical computer program.

There are many statistical computer programs or "packages" available. Almost all would be capable of the calculations covered in this book. However, none are ideal. Many are unnecessarily complex for use in straightforward situations. The complexity, profusion of options, and graphical output may serve to confuse and distract users interested only in straightforward situations. Many contain errors in that they use easy-to-program, approximate methods when exact methods are more appropriate. Some contain other errors. Few programs are available free of charge, even though most of the intellectual effort underpining such programs is ultimately a product of publicly funded universities in which academics have worked for the public good.

In response to these issues, I have written a statistical program to accompany this book. I have called the program "pds" for Public Domain Statistics. It is designed to run on the Windows operating system (version 95 or later), and occupies about 1 Mb. It is available for distribution free of charge with the proviso that it is not to be used against the interests of humanity and the environment. It is available on the World Wide Web at <http://www.jcu.edu.au/school/mathphys/mathstats/staff/DAKault.html>. It can also be obtained by personal request from the author at the Department of Mathematics & Statistics, James Cook University, Townsville, Qld 4811, Australia (please send the cost of postage and floppy disc). Source code can be made available to programmers who guarantee that extensions to this work remain within the public domain.

This book contains references to pds and brief instructions on its use, but the book can be used in conjunction with any other statistical computer program. Indeed, lack of access to a computer would be only a minimal handicap in using this book to gain an understanding of statistics.

CHAPTER 1

Statistics: The Science of Dealing with Variability and Uncertainty

Statistics can be defined as the science of dealing with variability and uncertainty. Almost all measurements made by scientists and people in many other fields are uncertain in some way. In particular, most measurement devices have limited accuracy, so there is uncertainty about the exact value. Sometimes what is being measured varies from individual to individual and from time to time, making it impossible to measure the true average exactly. For example, it is impossible to know exactly the true average blood pressure of the average healthy person.

Often we have to make a decision against a background of variability. We might be interested in a new blood pressure treatment. Should we believe that the new treatment works better than the standard treatment? The figures we collect after trying out the new treatment and comparing it with the standard treatment may slightly favor the new treatment. However, there is so much variability in blood pressure from person to person and from day to day that it will often be difficult to know whether it would be more reasonable to put slight changes in the average down to the effects of variability rather than to believe that the new treatment was superior to existing treatments.

THE QUEST FOR "OBJECTIVE" METHODS OF DEALING WITH UNCERTAINTY

By the early years of the twentieth century, the achievements of science had captured the public's imagination. There was a widespread desire to apply some scientific method to many areas of knowledge, to measure things, and to be "scientific" in how the measurements were interpreted. It no longer seemed good enough to simply look at a crop of wheat and note that on the side of the field treated with Bloggs's fertilizer the wheat grew better than on the side treated with Jones's fertilizer and conclude that Bloggs makes better fertilizer than Jones. Maybe on Bloggs's side of the field the soil was better to start with. More measurements were needed and these measurements had to be analyzed "scientifically." There was a need for a scientific approach to making decisions that took account of the variable nature of many types of measurements.

One important ingredient in the scientific approach seemed to be objectivity: Scientists were seen as using calculated reason based on hard facts. Making decisions based on guesswork and intuition did not seem to be part of the scientific method. There was therefore pressure to invent an objective method of drawing conclusions from uncertain or variable information. A method was wanted that did not depend on intuition. As a result, a method of objective decision making on the basis of variable data was developed early in the twentieth century and is widely used today. This method is properly called *frequentist statistics*. Other mathematically based methods of making decisions in the face of uncertainty were also developed and come under the heading statistics, but since these other methods are not objective and are often more difficult, they are not as well known. Frequentist statistics is so popular and so widely used that most people don't even realize that it is just one of a number of different varieties of statistics. For most people, frequentist statistics is "statistics." This book, too, will usually just use the word "statistics" in place of the mouthful "frequentist statistics."

"OBJECTIVITY": A MISTAKEN GOAL

Unfortunately, the pressure for an objective method of dealing with data was misguided. Most statisticians believe that the best ways of drawing conclusions in the presence of uncertainty involve methods that are not entirely objective. To give the appearance of objectivity, frequentist statistics starts with the premise that all decision making should flow from an analysis of the measurements that have been made. This approach has the added virtue that a computer program can be used to entirely automate the process of decision making. But this is often a ridiculous approach. We almost always know more about the topic than just the measurements, and surely it is silly to entirely ignore this knowledge in the decision-making process. Statistics, as currently used by most nonstatisticians, is the product of a mistaken quest for objectivity and simplicity.

STATISTICAL IMPERIALISM

Nevertheless, statistics has become the approach that the modern world takes to analyzing figures. Anybody who has to deal with making decisions on the basis of figures who simply looks at the figures rather than "get stats done on them," whether they be a researcher or an administrator, would be regarded as inadequate in their job and unable to cope with the intelligent, modern approach. Statistics has conquered the world of decision making. There is an almost religious belief that the modern world knows how to approach all problems and that "stats" is part of this approach. Nobody seems to notice that statisticians don't share this unthinking faith in statistics. Statisticians see some value in frequentist statistics, but many believe that it is not reasonable to try to deal with measurements involving uncertainty or variability in an entirely objective way. To base all analyses on figures alone means to abandon common sense, and often common sense can bring more wisdom to a subject than a blind analysis of figures. As a result of ignoring common sense, a considerable part of the world's scientific output is wasted effort. Analysis of experiments without the benefit of common sense can lead to misleading and sometimes dangerous conclusions. The use of common sense shows that many experiments should not have been performed in the first place.

A FIRST EXAMPLE OF THE CLASH BETWEEN STATISTICS AND COMMON SENSE

Let us look at an example where no great issues are at stake. We will consider two small groups of piano students. Say that in the first group the students got a half-hour lesson per week and in the second group the students got an hour lesson per week. Now assume that the results of the students in their piano exams showed that there was a lot of individual variation in ability but that the students who got the extra tuition time averaged out about 2 percent above the students who didn't. A normal person who had not had the "benefit" of a statistics "education" would conclude that the extra tuition time helped, but perhaps only a little. They might also think that perhaps the benefit may have been a bit underestimated due to a fault in the experiment; perhaps the experiment should have involved more students. By using common sense, the untrained person would come to appropriate conclusions.

On the other hand, someone who had been through a course in statistics as it is commonly taught would type the exam marks for both groups into a computer. A figure would come out of the computer. On the basis of this figure such a person would be likely to conclude that "there is no evidence of any benefit from the half hour extra tuition," or even worse, "statistical tests have proven that extra tuition is of no benefit." What the computer actually would have told them would be that "looking at the figures alone, the small difference of 2 percent between the groups could reasonably be attributed to individual variability causing the average of the second group to be a little higher just by chance." However, it is silly to look at the figures alone. We know that very few students will pass an exam with no tuition. Some tuition enables many students to pass. It seems reasonable to believe that additional tuition may enable many students to do even better. In other words, common sense tells us that, on average, extra tuition will almost certainly help students. Common sense here would lead to far better decision making than blind application of frequentist statistics.

THE LOGIC USED IN STATISTICS

Frequentist statistics can be valuable in decision making, provided it is not applied blindly. To apply statistics wisely requires an understanding of the rather convoluted logic that underlies frequentist statistics. It is the purpose of this book to show how statistics can be combined with common sense. The logic and philosophy of frequentist statistics are therefore fully explained in the next few paragraphs so that statistics can be used with common sense to make sensible decisions. Just a few minutes of concentration may be required for understanding. However, since the ideas can be awkward to follow, the explanation is repeated in the context of various examples throughout the book.

Let us look again at the example of the piano students receiving half an hour versus an hour of tuition per week. Although common sense tells us that the extra half hour of tuition will nearly certainly be of some help, the actual amount of benefit of 2 percent in exam marks in our figures turned out to be quite small. It is still just possible that students get all the tuition that they can absorb in one half-hour lesson each week, with the extra lesson being useless. If the extra tuition was in fact entirely useless, we could still account for the extra 2 percent marks in the extra tuition group by arguing that it was due to individual variability and that it was just coincidence that this variability turned out to favor the extra tuition group. In other words, it is just possible that the extra tuition was useless, but by sheer random chance there happened to be rather better students in the extra tuition group who got better marks, not because of the extra half hour, but because they were better students. It is therefore just possible that we are looking at a chance result that makes it appear that the extra tuition helps when in fact it does not.

Let us look at how a computer could help us here. Ideally, we would want to ask the computer, "What is the chance that the difference in the two groups of student pianists is not due to the benefit of the extra tuition but is instead due to individual variability just happening to favor the students in the group that got the extra tuition?" More generally, we often want to ask, "Is it reasonable to blame chance for the difference?" Any reasonable answer to such questions needs to take into account both the figures we obtain in our experiment and our common sense judgments. In the case of piano tuition, common sense tells us that it is exceedingly likely that the extra tuition will be of benefit. However, the computer can't take account of our common sense judgments since we are only telling it about the figures. Therefore, the computer can't answer the question, "Is it reasonable to blame chance for the difference?" Instead, it answers a related secondary question: "If the differences between two groups were entirely due to natural variability alone, how often would it turn out that the two groups end up at least as different as these two groups are?" Expressed another way, we want to ask, "Are the differences real or are they just due to coincidence?" Instead of answering that question we get a reply to the question, "If we were to put the differences down to sheer coincidence, what sort of coincidence would we be dealing with?"

The situation here can be compared with the situation of a biologist studying mammals on an island. The biologist may already know that there are cats on the island, but she may be interested in whether there are different mammals as well. Say the biologist came across some yellow fur. The biologist will want to ask, "This animal has yellow fur. Does this mean that there are different mammals here and not just cats?" In the analogy with the questions asked and the answers given by statistics, the biologist would receive an answer to the secondary question, "Do cats often have yellow fur?"

Question we want answered	Question actually answered
This has yellow fur, so is it a different mammal or is it just a cat?	Do cats often have yellow fur?
There is a sizeable difference between the two groups, so is there a real difference or is it just due to chance?	Does chance alone often lead to differences between two groups at least as large as the differences we see here?

Let us say that in the case of the exam results of the student pianists we got the reply from the computer, "Chance alone could often lead to a difference of 2 percent or more between the average of the two groups." From this, if we were just looking at the figures and ignoring any background knowledge or common sense, we could say, "The figures themselves give no convincing evidence of any benefit from the half hour extra tuition." However, it is traditional in frequentist statistics to leave out the important qualifiers "the figures themselves" and "convincing" from this sentence and instead state, "There is no evidence of any benefit from the half hour extra tuition." Even worse, this misleading form of words is sometimes further distorted to become "statistical tests have proven that extra tuition is of no benefit." However, as discussed, even if chance could easily account for the difference in the marks, it does not accord with common sense to say that we have proven that we should blame chance for the difference in the marks of the two groups of students.

Now let's say we got the opposite message from the computer: "Chance alone would hardly ever lead to such a big difference between the two groups." It is then traditional in frequentist statistics to make the decision that there is a real difference in the progress of the two groups of student pianists. What is meant by "hardly ever"? The actual result given by the computer is a probability. Traditionally, "hardly ever" is taken to mean less often than one in twenty times. The synonyms "p value less than 0.05," "statistically significant at the 0.05 level," "significant at the 0.05 level," or "statistically significant" are often used. If this happened in the case of the student pianists, we would be happy to agree with the conclusions reached by someone following the tradition of frequentist statistics. In other words, both common sense and frequentist statistics would tell us that we should believe that the extra tuition is of some benefit.

We have seen in the case of the student pianists an example where stats could tell us we shouldn't believe something makes a difference when common sense tells us that it does. There are cases where "stats" tells us that we should believe that something makes a difference, but common sense tells us that it doesn't. In such cases we make the judgment that chance, even a rather tiny chance, is a more reasonable explanation for the differences than the explanation that there is a real underlying difference. For example, let's say a friend claimed to be a clairvoyant. You tested her powers by seeing if she could guess some number between 1 and 100 that you had written on a piece of paper. If she happened to get the correct number and you were an unthinking frequentist statistician, you would now believe that your friend is a clairvoyant. Why? Because chance alone would hardly ever allow her to get the correct number. Here, the chance involved would be 1 chance in 100, or $p = 0.01$. Since this is less than a one-in-twenty chance, it is the sort of chance that hardly ever occurs, and so following the strict traditions of frequentist statistics we would say that there is statistically significant evidence that your friend is a clairvoyant. However, most people are at least a bit skeptical about clairvoyants, or at least won't readily believe that their friends are clairvoyants, and so most people would not be convinced by one correct guess out of 100 numbers. For these people, following the traditions of frequentist statistics would lead to a conclusion that they felt was not supported by the evidence. Some skeptics might want to see your friend correctly choose a nine-digit number—chance alone would allow a correct guess only once in a billion times—before they might start to believe that genuine powers of clairvoyance is a better explanation than chance. For such skeptics in this situation, the p value that is just tiny enough to make them change their minds would be one in a billion or $p = {}^1/_{1,000,000,000}$. Here it would be inappropriate for skeptics to use the traditional value $p = 0.05$ as the benchmark for the sort of chance that is unreasonably small. Instead, such skeptics should use $p = 0.000000001$ as the benchmark.

Restating, when we see a difference between two groups we might want to ask, "Could the difference just be due to coincidence?" Statistics does not answer this question, but instead answers the related question, "If we were to put the differences down to coincidence, what sort of coincidence would we be talking about?" The answer to this second question is the p value. If the p value is unreasonably small, smaller than some arbitrary benchmark (the coincidence is highly unlikely), it is more reasonable to believe the difference is "for real." There is a tradition of using a fixed value of 0.05 as the benchmark for what is unreasonably small. However, if this tradition is blindly accepted and the benchmark is not adopted to suit circumstances and common sense, "stats" can lead to unreasonable, even bizarre, decisions.

This section on the logic used in statistics requires some thought. The logic is a bit twisted and difficult. However, the ideas just explained are the main ideas underlying introductory statistics. If you understand these ideas, you have understood most of a first course in statistics. Because of its importance, the explanation will be repeated in various contexts throughout this book.

THE ROLE OF MATHEMATICS

In this book the understanding of statistics is enhanced by giving the complete mathematical basis for a few statistical tests. However, mathematical knowledge beyond the tenth-grade level is not assumed. Many statistical tests are derived using quite complex mathematics and involve a complex series of calculations. Other texts go to some length to detail all the mathematical manipulations that are required for these statistical tests. The attitude taken here is that if it is not possible to understand the mathematical derivation of the test, and if the details of the calculation don't help you to understand how the test works, then there is absolutely no point in learning the steps used in the calculations. Computers are now available to do these calculations. (A computer program called pds [public domain statistics] was written to accompany this book and is available free of charge for nonprofit purposes from <http://www.jcu.edu.au/school/mathphys/mathstats/staff/DAKault.html>.) The important issue is to understand the philosophy and assumptions behind the test. Therefore, if the details of the calculations are not enlightening in some way they are entirely omitted. As a result, there is less emphasis on calculation and formulas in this text than in most other introductory statistics books.

On the other hand, there are a few statistical tests for which the full derivation can be understood by anyone who can understand tenth-grade mathematics. These tests are explained in detail to enhance understanding of the nature of statistics.

THE REMAINDER OF THE BOOK

Before we return to the harder parts of statistics—using statistics to make decisions about whether we should believe that there are underlying differences between groups—we will briefly cover the easiest parts of statistics. Recall the definition of statistics: the science of dealing with variability and uncertainty. The first part of dealing with variability and uncertainty is to describe it. The part of statistics that deals simply with description is covered next. People who have any knowledge of statistics could just skim this chapter. The book then gives a simple introduction to the mathematics of chance: probability theory. This is necessary in order to understand some of the statistical tests that are described in full and in order to give you a feeling for situations that can reasonably be blamed on chance. The study of using statistics to make decisions occupies the remainder of the book.

In some places, interesting details and derivations are given that are not essential for people who just want the main ideas. Material such as this, which is redundant to the reader who does not want to cope with additional complexity, can be skipped without losing the main ideas of the book. Parts of the text containing such information are headed "Optional" and set between rules. In some places, these optional sections contain additional examples.

Answers to the questions at the ends of the chapters are given at the end of the book, beginning on page 241.

SUMMARY

- Statistics is a way of making a decision about whether differences are "for real" or just a result of chance.
- The form of statistics in common use attempts to be entirely objective and so just looks at the figures available and entirely ignores any common sense knowledge of the area.
- As a result, statistics cannot directly answer the question, "Is there a real difference?" Instead, it answers an indirectly related question: "If we were to blame coincidence for the difference, what sort of coincidence would we be talking about?" The answer to this question is called the p value. If the p value is smaller than some benchmark, the coincidence is regarded as unreasonably long and we conclude that there is a real difference.
- Traditionally, the benchmark p value is taken to be 0.05.
- Blind adherence to this traditional benchmark can lead to unreasonable decisions that defy common sense.

Put simply, the p value tells us how easy it is for chance alone to explain differences. It does not tell us how likely it is for chance to be the true explanation.

QUESTIONS

1. Think of another situation, like the music students and the extra tuition example, where you would believe that the benefits were "for real" regardless of results from a statistical analysis telling you that the favorable results could be easily explained by chance.

2. The roulette wheel in a casino can stop in thirty-eight different positions. The casino is known to operate the roulette wheel fairly. You notice that the roulette wheel hasn't stopped on "36" even once in the last 500 goes, so you place a bet on "36." What is the chance that you will lose your money?

3. Imagine your neighbor claimed to be a clairvoyant and asked you to verify her powers by getting you to write down a number between 1 and n (where n stands for a number like 10, 20, 100, or 500) that she then correctly guessed. How big would n have to be in order to convince you (assuming you can rule out cheating or magic tricks)?

4. Consider the same scenario as in question 3, but this time you are going to get the self-declared clairvoyant to perform the number guessing twice in a row and you will believe in her powers only if she is correct both times. Again, how big would n have to be in order to convince you (assuming you can rule out cheating or magic tricks)?

CHAPTER 2

Descriptive Statistics

NUMBER AND TYPE OF MEASUREMENTS

Generally, our first step in dealing with a situation in which uncertainty or variability plays a role is to make some measurements. The first question is how many measurements should we make. We could measure all the individuals in which we were interested, an appreciable proportion of them, or a negligible proportion of them.

Ideally, we would accurately measure all the individuals. This is called a census. Statistics would then just consist of describing the results in a digestible manner. Most times a census is not possible. Often the number of individuals we would need to measure is so large that it would not be possible to conduct a census with limited time and resources. For example, if we wanted to make a statement about the height of women that would always be correct we would have to measure all the women that have ever existed or that could ever exist in the future. Obviously, this is not possible.

When it is not feasible to measure all the individuals in which we are interested, we measure a selection of them. Usually the selection is a small or infinitesimal fraction of the number of individuals in which we are interested. In the case of women's heights, we would measure a small selection of women. The statistical term used here is that we take a "sample" from the population. The word "population" is used to describe the collection of all the objects we could measure, even when we are considering nonliving objects (e.g., the population of all possible midday temperatures). There are pitfalls in taking a sample. If we wanted to know about people's weights and we set up a weight-measuring facility outside the door of the Weight-Watchers Association, our sample of weights would obviously not fairly represent the weights in the population. For a sample to be fair, each member of the population has to have an equal chance of being chosen. A sample chosen this way is said to be a representative sample. A weight-measuring facility outside the door of the Weight-Watchers Association would not be representative because overweight people would have a greater than average chance of being chosen. There are many more subtle ways of obtaining a sample that is not representative of the population. Both the theory and the practicalities of obtaining representative samples are large areas of study in their own right. However, in most of what follows we will simply assume that we have obtained a representative sample. Much of statistics consists of calculations about how accurate information about a population is going to be when this information comes from a representative sample.

Occasionally the sample may consist of an appreciable proportion of the population, as in an opinion poll for the election of a mayor in a small town. Such an opinion poll might sample a quarter of the people who can vote. As well as simply representing the opinions in the whole population, this sample would also give us certain knowledge about an appreciable proportion of the population. As discussed in Chapter 10, some modifications to statistical calculations are then required. However, in most of what follows we will assume that our sample is a negligible fraction of the population and that it gives us ideas about the population in a probabilistic way.

For simplicity, instead of listing all the measurements in our sample we often want to describe them more briefly or express them in some summary form. The description or summary form can be in terms of summary numbers such as averages, or it can be in the form of diagrams. The type of summary that is used will depend partly on the type of measurements or data.

TYPES OF DATA

Measurements (the words "data" or "information" can be used interchangeably here) can be of three basic types: continuous, discrete, and categorical. Ordinal is a further type which is here described as a subtype of categorical data.

Continuous Data

In the case of continuous data the thing that is being measured can vary continuously. An example is the measurement of height. In practice, height measurements are often rounded to the nearest centimeter or millimeter, but it is possible for somebody to be, for example, anywhere between 172.3 cm and 172.4 cm in height. If we could use an infinite number of decimal places in measuring height, there would be an infinite number of heights distributed continuously between 172.3 and 172.4. Other examples of continuous measurement are weight, temperature, and ozone concentration. In the case of each of these measurements, the size of the step between one measurement and the next biggest measurement can be arbitrarily small, so there need be no cutoff in size between one measurement and the next. Since the size of the different measurements are not necessarily cut off from each other by any fixed amount, we regard the measurements as continuous data.

Discrete Data

In the case of discrete data, the measurement has to fall on separated (or discrete) values. Discrete measurements almost always arise from counting. An example is the number of accidents the clients of an insurance company have in one year. For any client the measurement can be 0 or 1 or 2 or 3 and so on. These numbers are discrete in that they are separated by whole units from each other. It would make no sense to have the number 1.3 as the number of accidents a client had. A second example of a discrete measurement is the number of children in a family, since children also come in whole numbers: Having 1 child or 2 children in a family makes sense, but no family contains 1.3 children.

Categorical and Ordinal Data

Data can also be categorical. This means that our measurement consists of simply classifying the object into one of several categories. An example is classifying people by their religion. Here our measurement is simply the classification of each individual as Christian, Buddhist, Moslem, and so on. A second example of a categorical measurement is to record the species of plants in a field. When there is only two categories possible, categorical data is sometimes called dichotomous data. Examples of pairs of dichotomous categories are yes–no, better–worse, and alive–dead.

There is another variant of categorical measurement. Categorical measurements are referred to as ordinal if the categories can be sensibly ordered. The different religions can't be ordered (except perhaps to some religious bigot it would not make sense to put Hinduism above or below Buddhism), so religion is not ordinal data. However, cancer patients can be ordered into those with stage I cancer, who have good survival prospects, and those with stage II, III, or IV, who have progressively poorer prospects. However, someone with stage II is not twice as badly off as a person with stage I or half as badly off as someone with stage IV. The numbers I, II, III, and IV make sense as an ordering, but not as numerical measure. This is the key feature of ordinal data. A second example of an ordinal measurement would be the classification of people into nonsmokers, light smokers, and heavy smokers. Smoking increases the risk of many diseases. Nonsmokers have less risk of these diseases than light smokers, and light smokers have less risk than heavy smokers, but it generally is not true that the risk for light smokers is halfway between the risk for nonsmokers and the risk for heavy smokers. The ordering nonsmoker, light smoker, heavy smoker is therefore useful, but it would not be useful to think of this ordering in the same way as we think of the numbers 1, 2, and 3, with 2 being exactly halfway between 1 and 3. A third example could be the classification of the age of animals as juvenile, immature, adult, or senescent.

Most writers regard ordinal data as a separate type from categorical data, so the common classification of data types are continuous, discrete, ordinal, and categorical. How data are classified also depends on our viewpoint. If we look at plants in a field one by one and decide which species each one belongs to, we are dealing with categorical data. If we look at the total number of plants of a particular species in the field, we can regard that number as one item of discrete data.

As well as classifying data as continuous, discrete, ordinal, or categorical, data can also be classified according to how many measurements are made on each individual. Where one measurement is made on each individual, the data are called univariate. If two measurements are made on each individual, the data are bivariate. Where more measurements are made on each individual, the data are multivariate. If we were just interested in height and measured the height of a number of people, we would have univariate data. If we were interested in describing the connection between height and weight and measured these two quantities on each of a number of individuals, we would have bivariate data. If we were interested in the connection between students' exam results and their home situation we might measure not only each student's exam results but also his or her parents' income, parents' educational achievements, number of siblings, number of hours of TV watched each night, and so on. This would be multivariate data.

SUMMARIZING CATEGORICAL AND ORDINAL DATA WITH NUMBERS AND DIAGRAMS

To summarize categorical and ordinal data with numbers we simply give a table listing the totals in each category. Such data can also be summarized with diagrams. Two sorts of diagrams are used: bar graphs and pie charts. These are just the names for the diagrams that many people will have seen in newspapers and elsewhere. As an example, say we picked out 100 adults at random and classified each person according to whether they were in the category "single," "married," "widowed," or "separated" or divorced." If the numbers in the various categories were 15, 45, 10, and 30, respectively, the table-form summary would be as follows:

single	married	widowed	separated/ divorced
15	45	10	30

The diagrammatic summary would be given by the following bar graph or pie chart. In the bar graph, the height of the bars gives the number in each category. In the pie chart, the percentage of people in each group is mirrored by the size of each "slice" as a percentage of the entire "pie":

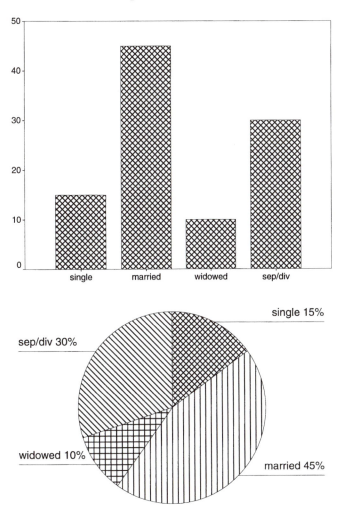

Exactly the same methods of display apply to ordinal data. If the figures 15, 45, 10, and 30 referred instead to the number of people with stage I, II, III, and IV cancer, replacement of the labels single, married, widowed, and sep/div by I, II, III, and IV is all that is needed. However, people concerned that diagrams should obey a strictly logical layout would prefer a bar graph rather than a pie chart representation for ordinal data. In a bar graph the bars could represent people with stage I, II, III, and IV cancer in that order; whereas in a pie chart, stage IV cancer would be displayed as lying between stage III and stage I, not a logical position for it to be in.

SUMMARIZING DISCRETE AND CONTINUOUS DATA WITH NUMBERS

The most obvious method of summarizing a list of measurements of height or weight or whatever is to calculate the average. Why then the need for more discussion? There are two answers. First, taking the average is not the only method of describing where the data are centered. There are other methods of describing where the data are centered that in some circumstances give more insight than just quoting the average. Second, in addition to a summary measurement telling us where the data are centered, we may want a measure of how spread out the data are. Summary measures of where data are centered and how spread out data are are known as *summary statistics*.

Where Are the Data Centered?

There are three measures of where data are centered (otherwise referred to as measurements of central tendency). These are the mean, median, and mode.

The Mean

The mean is simply a fancy term for average as learned in primary school. Simply add up all the values and divide by the number of values. For example, the mean of 11, 10, and 21 is

$$14 = \frac{11 + 10 + 21}{3}.$$

The mean of a sample, or sample mean, is often given the symbol $\bar{x}$.

OPTIONAL

It is useful to write rules, such as the rule for finding the mean, in a way that does not depend on examples. However, this requires getting used to a bit of mathematical jargon. To spell out the rule for finding the mean using the correct mathematical jargon, we have to replace the numbers in the example with symbols. For the first data value we use the symbol x_1, for the second x_2, for the third x_3. In our example, $x_1 = 11, x_2 = 10, x_3 = 21$. If there were more values, we would have an x_4, x_5, and so on. If there were n values, our last value would be x_n. If we want to refer to one particular data value but we don't want to be specific that we are dealing with the first, second, third, or nth value, we use the symbol x_i. If we then said that i was 1, we would be dealing with x_1, if i was 2, we would be dealing with x_2, and so on. To find the average, we first add all the data values. In symbols, this is $x_1 + x_2 + x_3 + \ldots + x_n$ where the $+ \ldots +$ here means to keep on adding up all the values in between the third and the nth.

Adding all the data values can be more compactly written as

$$\sum_{i=1}^{n} x_i.$$

The symbol Σ is the Greek equivalent of the English "S" and stands for "sum up." The symbolism literally means sum up all the values of x_i where i takes each of the values in turn between 1 and n; in other words, $x_1 + x_2 + x_3 + \ldots + x_n$. With this notation, the formula for finding the mean is written

$$\frac{\sum_{i=1}^{n} x_i}{n}$$

so we write

$$\overline{x} = \frac{\sum_{i=1}^{n} x_i}{n}$$

END OPTIONAL

If we have measured all the individuals in the population, we can calculate the true population mean. The formula is the same as for the sample mean, but we use the symbol μ (the Greek letter mu) instead of $\overline{x}$. In general, numbers describing a population are denoted by a Greek letter and estimates of these numbers obtained from a sample are denoted by an English letter.

The Median

The median is obtained by writing the data in ascending order and finding what the middle value is (in our previous example with the numbers 11, 10, and 21, the numbers in ascending order are 10, 11, 21). The middle value (here it is 11) is called the median. In other words, the median is halfway along the data from smallest to largest. If the data values were 13, 2, 7, 19, 154, 26, and 38, the median would be 19.

OPTIONAL

If there are an odd number of data values, it is always possible to find a middle number. If there are $2n + 1$ data values, the $(n + 1)$th has n values below it and *n* values above it. If there are an even number of data values, the median is taken to be the average of the two numbers closest to the middle. Say the data values were 13, 2, 7, 19, 154, and 26; the median is taken to be 16, which is the average of 13 and 19, since there are three numbers below 16 and three numbers above 16 and 16 is halfway between the two middle numbers, 13 and 19.

END OPTIONAL

The Mode

The mode is the most commonly occurring value. If the data values consisted of the numbers 10, 10, 11, and 21, then the mode would be 10. If the data values were 19, 20, 17, 21, 20, 18, 20, 154, 19, and 756, the mode would be 20. If the data values were 19, 20, 17, 21, 756, 20, 18, 20, 154, 19, 756, and 756, then there would be two modes: 20 and 756. The data in this case are said to be bimodal. In contrast, data that have just one mode are referred to as unimodal.

OPTIONAL

The term bimodal is also used, loosely, to describe situations where there are two different values that are more common than neighboring values but not necessarily just as common as each other. For example, completely clear days are more common than days with 1 percent cloud, 2 percent cloud, and so on, and completely overcast days are more common than days with 99 percent cloud, 98 percent cloud, and so on. In this situation we use the term bimodal to describe cloud-cover data. Strictly speaking, the data have two modes only if the most common values, 0 percent and 100 percent, occur just as often as each other. Nevertheless, the cloud data we have described would be called bimodal even though 0-percent cloud days are not exactly as common as 100-percent cloud days.

END OPTIONAL

The mode is generally useful when the data are discrete and the number of values that are likely to occur are fairly small. For example, the mode is a useful measure of the central tendency in family size. The number of children in families can take only a relatively small number of discrete values, say 0, 1, 2, 3, . . . , 20. If families with one child are more common than families with any other number of children, we say the modal number of children is 1. As a measure of the center of the data it could seem more reasonable to use this mode of 1 than to use the mean number of children, which might be 1.83. The mean is a valid measure of central tendency, but it doesn't give us an immediate picture of a typical family the way the mode does. The mode would similarly be useful if we were talking about weekly earnings rounded to the nearest $100. Only the numbers $200, $300, . . . , $2,000 are likely to be particularly common. Numbers like $1,328,400 for weekly earnings, while they may occur, would be very rare. However, this example leads us to the disadvantages of the mode. What if we wanted to get the mode more accurately and so asked people their weekly earnings specified to the exact number of cents? Then we might find that even in a large survey just about everybody was different, at least in the cents. If everybody's income was different, this would make everybody's income a mode (each value would occur just as commonly as every other value), and then the mode wouldn't be very useful. There wouldn't be a single mode that would stand out as the center of the data. In general, it is not useful to try to precisely locate the mode when dealing with continuous data or when dealing with discrete data where the units are tiny (e.g., cents) compared to the data values (e.g., hundreds of dollars). As a result, in most situations modes are not a commonly used measure of central tendency.

Reasons for Different Measures of Central Tendency

We've just discussed the advantages and disadvantages of modes compared to means as a measure of central tendency. What are the advantages and disadvantages of medians compared to means? Which is best? The answer to this question is that there is no absolute right way of defining where data values are centered. Different definitions have different purposes. Consider income for adults in the United States. Very roughly, there are about 200 million adults and the combined weekly income is about $400 billion. Dividing 200 million into $400 billion gives us a mean weekly income of $2,000. Most people will protest about such a figure being declared a measure of where weekly income in the United States is centered. Most people would say that they know hardly anybody who earns that much. If we were to round income to the nearest $100 per week, we might find that $500 per week is the most common or modal income, but, as discussed, it might not be meaningful to try to define the mode much more precisely. Why is this rough mode of $500 so different from the mean of $2,000? After all, both are meant to be measuring where the data are centered. The answer to why mean and mode can give very different impressions of where the data are centered can be understood by drawing a horizontal line on a page. The left-hand end of the line could be marked 0 to stand for a weekly income of $0, 1 cm to the right denotes an income of $100 per week, 2 cm to the right denotes $200 per week, and so on. Let's imagine we make marks on the line corresponding to each person's weekly income. What we are making is called a line graph.

236 meters up the road

Line graph showing income of a few of the millions who earn under $1,000 per week and one "high flyer." Axis labelled in centimeters, with each centimeter representing $100 per week.

Chances are that just about everybody you and I know would have their income marked somewhere along the first 40 cm. However, there are a handful of people with enormous incomes of several million dollars per week. On our line graph these people would be marked hundreds of meters off the right of the paper. Those data values that are remarkably extreme are referred to as "outliers." These sort of data, where values trail out particularly in one direction, are referred to as "skewed data." In particular, we would describe income data as being markedly skewed to the right or as having a long right-hand tail. The average of a few measures of hundreds of meters, perhaps a few hundred measures of tens of meters, a few thousand measures in the meters, along with millions of measures of around 5 cm turns out to be about 20 cm. The average or mean income is likewise about $2,000, but this measure is not an accurate reflection of the financial standards of the vast bulk of people.

In this situation, whether we regard the mean or the mode as the best measure of where income is centered is really a political decision. If we are interested in the living standards of typical members of the society, the mode is a better measure. However, politicians with a vested interest in making figures look good or with an ideology that values wealth rather than people might regard the mean as a measure that better reflects their purposes. On the other hand, those who want to use the mode face the disadvantage of not usually being able to define it precisely. The compromise choice is the median. By definition, half the population earns below and half above the median, but the extreme incomes of the ultrarich do not influence the position of the median. In general, with skewed data the median will lie between the mode (to the extent that this can be estimated) and the mode. Medians are often regarded as a fairer measure of the center of skewed data than means. However, means are generally easier to deal with mathematically and are generally regarded as the fairest measure of where data are centered when the data are not particularly skewed.

Measures of the Variability, Spread, or "Dispersion" of the Data

Just as with measures of where the data are centered, there are several measures of how spread out the data are, each with its own advantages and disadvantages.

Range

The most obvious measure of spread is the range, which is the gap between the highest and lowest values. This measure is fine, provided we just want to summarize the spread of the data we have at hand. However, often we are interested in what the data tell us about the spread of values in the population. Say we wanted information on the spread of heights of adult women. If our sample consisted of just one woman, the smallest value we had in our sample and the largest value we had in our sample would be the same thing, the height of the one woman we had chosen. In this case the range would be 0 cm. If just a handful of women are chosen, chances are that all will have a height between 1.50 m and 1.75 m, giving a range of 25 cm or less. However, if a large number of women are examined, sooner or later we will come across a woman who is exceptionally tall and sooner or later we will also come across a woman affected by dwarfism. When we have collected enough data to include such exceptional people, the range may turn out to be 75 cm or more. The range then reflects how big our sample is as well as reflecting how spread out heights are in the population. It would be preferable to find a measure of spread that reflects only the spread of the values in the population and that isn't greatly influenced by the number in the sample. For this reason, the range is not generally regarded as a satisfactory measure of spread.

Mean Deviation

There are several approaches to finding a measure of spread that is not substantially affected by the number of data values obtained. One approach is to find the average of the spread of the data about the mean. This is called the mean deviation. Say we measured the heights of just three women and they were 159 cm, 165 cm, and 171 cm. The mean is 165 cm. Two measures are 6 cm from the mean and the other is 0 cm from the mean. We say that the measures deviate by 6 cm, 0 cm, and 6 cm from the mean. The average of 6, 0, and 6 is 4. This could be our measure of spread in this case. Clearly a procedure like this could be generalized to a case when there are more data values, and the average of the deviations from the mean could be used as a measure of spread. There is little wrong with this measure of spread, but it is not used because of mathematical difficulties. In particular, the value 171 is 6 above the mean and the value 159 is 6 below the mean. The average of 6 above (+6), 6 below (–6), and 0 is 0. In general, since the average is in the middle, there will always be values as much above the average as there are below the average, so the average of the deviations will always be 0 unless we ignore the minus signs that are attached to some of the deviations. The act of ignoring minus signs is referred to as taking absolute values. The average of the absolute values of the deviations is then a possible measure of spread. However, absolute values are awkward to deal with mathematically.

OPTIONAL ▰▰

Mathematics used in the theory of statistics in effect consists of rules to find best answers partly by using a method (calculus) that follows trends. An abrupt change of rule so that numbers slightly above 0 keep their correct sign but numbers slightly below 0 have their sign reversed, makes for difficulties in following trends.

▰▰▰▰▰▰▰▰▰▰▰▰▰▰▰▰▰▰▰▰▰▰▰▰▰▰▰▰▰▰▰▰▰▰▰▰ *END OPTIONAL*

Therefore, the mean deviation is little used as a measure of spread.

Variance

Squares of numbers are much more convenient mathematically than absolute values.

OPTIONAL ▬▬▬▬▬▬▬▬▬▬▬▬▬▬▬▬▬▬▬▬▬▬▬▬▬▬▬▬▬▬▬▬▬▬▬▬

In terms of calculus, the function $y = x^2$ is differentiable—that is, it is smooth—whereas the function y = the absolute value of x is not differentiable. The graph of this function comes to a sharp point at the origin.

▬▬▬▬▬▬▬▬▬▬▬▬▬▬▬▬▬▬▬▬▬▬▬▬▬▬▬▬▬▬▬▬▬▬ ***END OPTIONAL***

Like absolute values, squares also turn a mixture of negative values such as –6 and positive values such as 6 into all positive values (–6 × –6 = +36). Instead of averaging the absolute values of the deviations from the mean, one of the measures of spread commonly used is the average of the squared deviations. This measure is called the variance.

OPTIONAL ▬▬▬▬▬▬▬▬▬▬▬▬▬▬▬▬▬▬▬▬▬▬▬▬▬▬▬▬▬▬▬▬▬▬▬▬

The variance has a nice mathematical property. In particular, it can be shown that under some common circumstances variability as measured by variance adds up in a straightforward way when variable values are added. For example, if we knew that the variability of the heights of pony backs as measured by variance was 225 units and we knew that the variability of the heights of women as measured by variance was 100 units, then under some common circumstances the variability of the distance from the ground to the tops of the heads of women standing on pony backs would be 325 (225 + 100). This addition of variability as measured by variance applies whenever there is no tendency to pair tall women with tall ponies or vice versa, a condition known as independence. By contrast, the heights of women and the heights of their husbands are unlikely to be independent, as tall women tend to prefer to marry even taller men. In this case we would say that the heights of women and the heights of their husbands are dependent.[1]

▬▬▬▬▬▬▬▬▬▬▬▬▬▬▬▬▬▬▬▬▬▬▬▬▬▬▬▬▬▬▬▬▬▬ ***END OPTIONAL***

Standard Deviation

Variance as a measure of variability has a major drawback. Because of the squaring, the units of measurement don't match up with what has been measured. The height of women might be measured in cm, but the variance would then be in square cm. To get back to the original units we need to take a square root. The square root of the variance is the most widely used measure of the spread of data. It is called the standard deviation. The standard deviation is then the square root of the average of the squared deviations. The standard deviation of a sample is denoted s; the standard deviation of a population is denoted by the Greek letter σ.

OPTIONAL

There is, however, one further minor modification required in the definition of the standard deviation. If we had only one data value, the average would be the same as the data value. The deviation would then be zero and the square root of the average of the squared deviations would also be 0. The standard deviation, then, as we have described it, would be zero

$$\left(\sqrt{\frac{0^2}{1}} = 0 \right)$$

whenever we have just one data value. But this is silly. If we have only one data value, we cannot estimate spread. We shouldn't say pony heights have zero standard deviation (i.e., don't vary at all) simply because we have only measured one pony. Any formula for calculating spread when there is only one data value should give the answer "undefined"

$$(\text{e.g.}, \frac{0^2}{0}), \quad \text{not zero } (\frac{0^2}{1}).$$

We deal with this difficulty by modifying our method of calculating the average of the squared deviations. We sum all these squared deviations but divide by one less than the number of squared deviations instead of dividing by the number of squared deviations. Put another way, if there are two data values, we really have information about only one deviation even though we can define two deviations from the average. Likewise, if there are n data values, in a sense we have only $n - 1$ measures of the deviations. For this reason, in calculating the average of the squared deviations we should divide by $n - 1$ rather than n. The formula, then, for the standard deviation of a sample is

$$s = \sqrt{\frac{\sum_{i=1}^{n} (x_i - \bar{x})^2}{n - 1}}$$

If we have measures on the whole population, a census, we calculate the population standard deviation using the formula

$$\sigma = \sqrt{\dfrac{\displaystyle\sum_{i=1}^{n}(x_i - \mu)^2}{n}},$$

where μ is the population mean. It can be shown that in some sense the formula for s involving division by $n-1$ is the "best" estimate of the population standard deviation σ (see Chapter 4). When n is large there will be a negligible difference between the results of using the formulae for s and for σ.

Although standard deviations have the advantage of the same units of measurement as the original values, one nice mathematical property does not hold. Unlike the situation with variances, standard deviations don't add in a straightforward way. Referring to the example of the heights of ponies and women in the discussion about variance, if the standard deviation of the heights of pony backs is 15 cm ($\sqrt{225}$) and the standard deviation of the heights of women is 10 cm ($\sqrt{100}$), the standard deviation of the heights of women standing on pony backs will not be $15 + 10 = 25$ cm. Instead, if we want to find the standard deviation of the heights of pony–woman combinations we need to use the fact that variances add. We know that the variance of the heights of pony–woman combinations is $225 + 100$, so the standard deviation will be $\sqrt{225 + 100}$, or about 18.03 cm (assuming independence). The fact that standard deviations don't add, unlike their squares (variances), is a reflection of the general mathematical rule $\sqrt{a^2 + b^2} \neq a + b$.

END OPTIONAL

Interquartile Range

Recall the disadvantage of the mean in dealing with data with a long tail, such as income data. If we use the mean as a measure of central tendency, a few extreme high fliers gives us a distorted view of where most people are at financially. The same sort of disadvantage applies when using standard deviation as a measure of spread. With the standard deviation, a few extreme deviations give us a distorted view of the size of typical deviations. The interquartile range (IQR) is a method of measuring spread that is not unduly sensitive to a few extreme deviations. The interquartile range can be roughly described as the range of values that contains the middle half of the data, with a quarter of the data values being below the interquartile range and a quarter of the data values being above the interquartile range.

To be more precise requires a definition of quartiles and percentiles. The first quartile is one-quarter of the way along the data from smallest to largest. Likewise, the third quartile is three-quarters of the way along the data from smallest to largest. The first quartile can be called the twenty-fifth percentile. Other percentiles are defined similarly. The median could be called the second quartile or fiftieth percentile. The interquartile range is then the gap between the first and third quartiles. For example, if the data consisted of the numbers 1, 2, 2, 7, 8, 9, 9, and 24, the first quartile would be 2, the third quartile would be 9, and the interquartile range would be 7. Because of mathematical difficulties, the interquartile range is less widely used than standard deviation.

OPTIONAL

There are some minor difficulties in deciding exactly which point should be regarded as one-quarter of the way through the ordered data. Unless the number of data points is a multiple of 4, it is not possible to divide the ordered data exactly into quarters. We can define the first quartile regardless of whether the number of data points is an exact multiple of 4 by finding a data value so that counting this particular value and those below it we have one-quarter or more of all the data values. Yet counting this particular value and those above it we have three-quarters or more of all the data values. It is possible for there to be two particular values that satisfy this property (this occurs when the number of data points is in fact a multiple of 4; if there are $4n$ values, both the nth and the $(n + 1)^{\text{th}}$ satisfy the property). In this case, we take the average of these two values. The third quartile and the various percentiles are defined similarly.

The mathematical difficulties that restrict the use of the interquartile range reflect the fact that the rule for averaging squared deviations and taking the square root to give the standard deviation can be stated reasonably simply, whereas obtaining the interquartile range involves the more complicated process of comparing each data value with all others to identify the values one-quarter way through and three-quarters way through the ordered data. Mathematics is needed here to work out the amount of variability likely when different samples from the same population are the source of the summary statistics.

END OPTIONAL

Table 2.1 provides a list of the numbers used to summarize data.

SUMMARIZING DISCRETE AND CONTINUOUS DATA WITH DIAGRAMS

As with summarizing data by numbers, there is no single method of summarizing and displaying data by diagram. Different methods have different advantages and disadvantages.

Line Graphs

Line graphs, where the location of each data point is marked along a line, were mentioned in the discussion of the mode. A line graph was used to dis-

Table 2.1
Numbers Used to Summarize Data

Meaures of Central Tendency	Definition	Symbol	Advantages	Disadvantages
Mean	Simple average	$\bar{x}$ (μ)	mathematically convenient	unfair representation of center of highly skewed data
Median	Half way along ordered data		more fair representation of center of highly skewed data	mathematically inconvenient
Mode	Most common value		more fair representation of center of highly skewed discrete data	not meaningful when data takes continuous or finely divided discrete values
Measures of Spread				
Range	Gap between smallest and largest		Simple	Reflects size of the sample as well as the underlying spread in the population
Variance	average of the squares of the deviations from the mean (divide by n - 1 rather than n in taking the average)	var	Mathematically convenient. Additive property (if there is independence)	Unit of measure is the square of data units. Unduly affected by a few large deviations
Standard Deviation	square root of variance	s (σ)	Mathematically convenient	Unduly affected by a few large deviations
Interquartile Range	Range of values containing the middle half of the data	IQR	Not unduly affected by a few large deviations	Mathematically inconvenient

play hypothetical data on income distribution on page 17. This is the simplest form of diagrammatic representation of data, though it is not widely used. It has the disadvantage that the diagram becomes unclear when the number of data points is large. Line graphs also can't display repeated instances of exactly the same values. This is a particular problem in the case of discrete data.

Stem and Leaf Plots

A stem and leaf plot is a way of arranging the printing of data values on the page so that the overall shape of the print gives an impression of the shape of the data. It has another use, in being a convenient method of sorting data by hand from smallest to largest. It has the disadvantage of only being able to display data to two significant figures (e.g., the tens column and the units column).

The following example explains the idea. Say some pollution measure on twenty-five different days were 31, 37, 22, 4, 19, 42, 34, 28, 51, 11, 27, 32, 47, 34, 15, 26, 18, 30, 43, 26, 44, 30, 22, 13, and 39. We first see that all data values, written in the form of x tens and y units, have an x value of between 0 and 5 (i.e., all the data values are between the units and the fifties). We then write the numbers 0 to 5 down the page, denoting the possible tens column values, followed by a vertical line. These numbers are the stems. We then look through the data values one by one and write the unit value for each data item adjacent to the appropriate tens value, continuing until all the items have been written down. These unit values are the leaves (see Figure 2.1).

Finally, we order the unit numbers to give the completed stem and leaf plot. It can be seen that the stem and leaf plot displays all the data that were collected to two significant figures, but displays the data in a form that gives a visual impression like a histogram on its side (see Figure 2.2).

Histograms and Ordinate Diagrams

Histograms are the most widely used way of displaying continuous and discrete data in diagrammatic form. The first step in drawing a histogram is the preparation of a frequency table. The following frequency table is based on the same data as in the stem and leaf plot. It can be obtained by looking through all twenty-five data values and noting that there is just one value between 0 and 9 (inclusive), five values between 10 and 19 (inclusive), and so on. It can be obtained more easily from the stem and leaf plot by counting the number of leaves for each of the stems(see Figures 2.3 and 2.4).

Histograms represent numbers by area rather than by height. Usually interval widths are equal (i.e., the widths 0–10, 10–20, and so on in the histogram here are equal). With equal widths, areas and heights are proportional and we can think of the histogram as representing numbers by height. However, in unusual cases we may want histograms where the interval widths are unequal. For example, say we had the information that of 2,000 newborns who died, 1,000 did so within twenty-four hours of birth and the remaining 1,000 died between twenty-four hours and thirty-one days. The death rate per day after twenty-four hours is one-thirtieth the death rate in the first twenty-four hours. The histogram should then consist of a column one day wide and 1,000 units high followed by a column thirty days wide and one-thirtieth the height of the first column. The two columns will have equal areas, since there are 1,000 deaths in each group (see Figure 2.5).

Figure 2.1
Stem and Leaf Plot

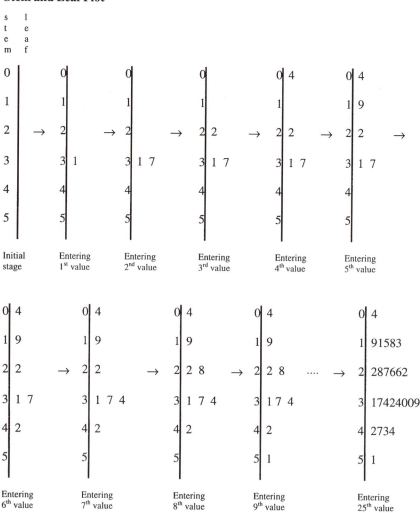

The boundaries of the columns in a histogram are another source of possible confusion. If we are dealing with continuous data, a figure of 30 will be a figure rounded to two significant figures. The figure 30 therefore represents a true value somewhere between 29.5 and 30.5. We see that any value between 29.5 and 39.5 would therefore fall into the interval labelled 30 to 40 on the histogram presented in Figure 2.4. The boundaries of the intervals should therefore be labelled 29.5, 39.5, and so on. The labelling of this histogram is therefore in error by 0.5. There is one exception. Ages are generally rounded down, not simply to the closest whole number. A person aged forty-nine years

Figure 2.2

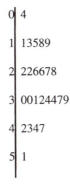

```
0  4

1  13589

2  226678

3  00124479

4  2347

5  1
```

Figure 2.3

Range	0-9	10-19	20-29	30-39	40-49	50-59
Number of values	1	5	6	8	4	1

Figure 2.4

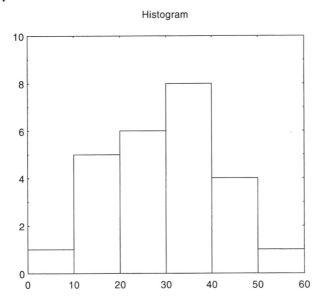

Histogram

and eleven months will tend to give his or her age as forty-nine, not fifty. The existing labelling on the histogram is thus valid for age data.

The last point about histograms is that they are not a strictly logical representation of discrete data. If we drew a histogram for the number of children

Figure 2.5

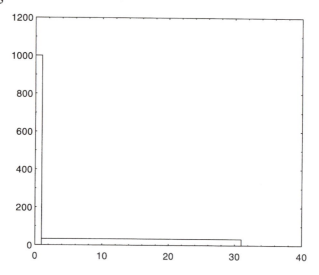

in a family, the fact that, say, twenty-three families had two children would be represented by an area 23 units high sitting on a base between 1.5 and 2.5. But this is a rather silly representation: Values between 1.5 and 2.5 other than the value of exactly 2 are not possible. The response to this complaint is to draw what is called an ordinate diagram, rather than a histogram. Instead of using an area of some width to represent the number in the category, a vertical line or ordinate is used and its height represents the number in the category (see Figure 2.6).

Figure 2.6

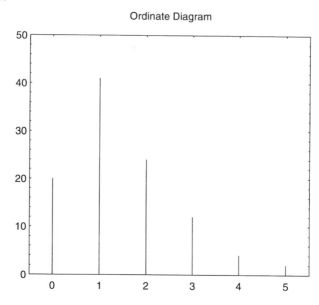

Boxplots

Boxplots are diagrams that show the interquartile range and median of the data as a rectangular box with a line at the median drawn across the box. "Whiskers" extend from either end of the box to display the range of data values. There is only one logical way of drawing a histogram, but conventions regarding the display of the range of values on boxplots are arbitrary and vary from author to author. Usually the whiskers extend to the furthest data points within a distance of 1 or 1.5 IQR on either side of the box with any more extreme points (outliers) being marked individually. The boxplot presented here represents the same data as in the stem and leaf plot and the first histogram (Figure 2.7).

Boxplots generally display less information than histograms. A histogram with many columns will give a detailed picture of the location of the data values to within the width of a narrow column, whereas a boxplot does little more than show a division of the data into four quartiles. However, boxplots are useful for making a large number of visual comparisons. Imagine that we wanted to compare peoples' incomes from twenty different regions. A set of twenty histograms to display all the data could not be easily absorbed by the eye. However, twenty boxplots could be drawn, one underneath the other down the page, and it would be obvious which region had larger overall incomes and which region had the greatest amount of inequality in terms of the spread of values between rich and poor.

Figure 2.7

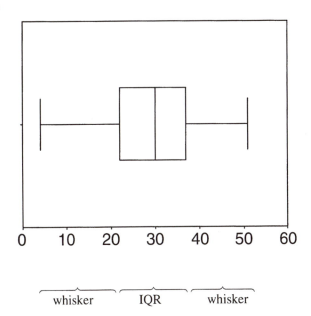

SUMMARY

- Most populations are large or infinite and a census is not feasible. Instead, we select a sample of a relatively small number of individuals to give us probabilistic ideas about the population.
- There are pitfalls in trying to select a sample that is representative.
- Data or measurements are of various types: categorical (including dichotomous), ordinal, discrete, and continuous.
- Different data types are summarized in different ways.
- Categorical and ordinal data are easily summarized by tables, bar graphs, and pie charts.
- Discrete and continuous data are summarized pictorially by stem and leaf plots, histograms, and boxplots.
- Discrete and continuous data are summarized numerically by measures of central tendency and measures of spread, as listed in Table 2.1.

QUESTIONS

1. Measure the heights of the people in your class. Make a stem and leaf plot to help you both order and display the data. Then create a frequency table, a histogram, and a boxplot to display your results (for the boxplot you will need to specify the rules you use in its construction; in particular, what rule determines the length of the whiskers?). Calculate the mean and standard deviation (you will have already calculated the median and interquartile range as a preliminary to drawing the boxplot). Note that some or all of the work can be done by a statistical program on a computer (although the pds program, written to accompany this book, does not do diagrams). However, hand calculations to enhance understanding and to verify the accuracy of the computer program are worthwhile initially. Computer calculations may be affected by programming error. Furthermore, the computer may not always do the calculation that you expect it to do, for various reasons: The data may not be in the format the program expects, you may press the wrong button, or the computer may be programmed to base its calculations on slightly different definitions from the ones taught here (which is likely in the case of interquartile range).

2. An opinion poll is taken by phoning people whose names are chosen at random from the telephone book. Why might this sample not be representative?

NOTE

1. This additive property is proven using the definition of variance as the average of the squared deviations, and some algebra. The rough outline of the proof follows. If the height of the ith woman is x_i and the height of the jth pony is y_j, then the deviation of the height of the ith woman and j^{th} pony combination from the average is $[(x_i + y_j) - (\bar{x} + \bar{y})]$. From the definition, proving that variances add, amounts to proving that the average of

the square of deviations, $[(x_i + y_j) - (\bar{x} + \bar{y})]^2$, over the various combinations of values of i and j is the same as the sum of the averages of $(x_i - \bar{x})^2$ and $(y_j - \bar{y})^2$. The algebra involves writing $[(x_i + y_j) - (\bar{x} + \bar{y})]^2$ as $[(x_i - \bar{x}) + (y_j - \bar{y})]^2$ and expanding this term, but there are complications in the algebra here. The expansion gives cross terms, just as the expansion of $(a + b)^2$ gives the cross term $2ab$ as well as the more obvious terms $a^2 + b^2$. However, providing we are dealing with all possible combinations of values of i and j, the cross terms turn out to be zero for the same reason that the average of all positive and negative deviations about the average are zero.

CHAPTER 3

Basic Probability and Fisher's Exact Test

To deal further with the variability and uncertainty that commonly arise whenever measurements are made, we need to know a little about the mathematical theory of chance: probability. In this chapter we will cover the basic ideas of probability and use the ideas to explain our first statistical test: Fisher's exact test.

DEFINING PROBABILITY

There are at least three approaches to defining probability, all of which turn out to be compatible with each other:

1. In terms of the long-run proportion of the time that something happens. For example, the proportion of time "6" appears on throwing a fair die many times is one-sixth. We say the probability of throwing a "6" is one-sixth. The expressions "the proportion of the time something happens" and "the probability that something happens" are taken to be equivalent.

2. In terms of a measure of belief. For example, we can state a subjective probability about the chance of a nuclear war over the next ten years. A person might believe that this probability is one-sixth. This would be a valid statement of belief, even though it is clearly not possible to view thousands of worlds just like our own and observe that one-sixth of them blow themselves up with nuclear weapons over the course of ten years.

3. As an abstract mathematical construct from basic rules. The rules say that probability is a number between 0 and 1 that can be attached to any event that might occur. A probability of 0 means the event won't occur. A probability of 1 means the event will occur with certainty. The probabilities of events that can't both occur at the same time add in the ordinary way to give the probability of either event occurring. For example, the probability of a die landing on either a "5" or a "6" is $1/6 + 1/6 = 1/3$.

As simple as these axioms are, they give rise to a large mathematical theory that can be shown to be compatible with both the "proportion of the time" and "measure of belief" approach to probability, but the details of this are beyond the scope of this book.

PROBABILITIES OF COMBINATIONS OF EVENTS

Often it is necessary to deal with the probabilities of combinations of events. In dealing with combinations of events, it is sometimes appropriate to add their probabilities and sometimes appropriate to multiply them. The rules about finding the probabilities of combinations of events can be derived mathematically from the axioms. We will instead just quote the rules, but we will also show that the rules are sensible by using examples and diagrams.

Addition Rule for Probabilities: "or"

In situations where we want to know the probability that either one event *or* another particular event has occurred but where the two events can't occur together, we simply add the probabilities (this is in fact one of the axioms of mathematical probability theory). For example, if the probability that a person has one brother is 0.4 and the probability that a person has no brothers is 0.3, then the probability that a person has no more than one brother is $0.4 + 0.3 = 0.7$.

There is a pictorial way of describing events and their probabilities and the way they combine: Venn diagrams. In these diagrams events are represented by patches in a rectangle, with the size of each patch being roughly proportional to its chance of occurring.

The Venn diagram for the addition rule is shown here, where A is the event a person has no brothers and B is the event a person has exactly one brother. The diagram shows no area of overlap between the areas covered by A and B, for it is impossible for both to be true at the same time. We say the events in these circumstances are mutually exclusive. The unlabelled area in the diagram represents the event that neither the event "has no brothers" nor the event "has one brother" is true. In other words, it represents the event that the person has more than one brother.

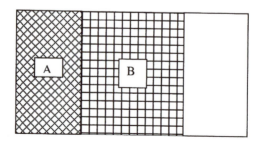

The addition rule can be extended to the case of more than two mutually exclusive events. In our example, if we were told that the probability of having exactly two brothers was 0.1 and that the probability of having exactly three was 0.08, then the probability of having three or fewer would be 0.4 + 0.3 + 0.1 + 0.08 = 0.88.

Multiplication Rule for Probabilities: "and"

In situations where we want to know the probability that both one event *and* another particular event have occurred, and where the occurrence of one event does not affect the occurrence of the other event, we simply multiply the probabilities. For example, if the probability that a person has one brother is 0.3 and there is a probability of 0.2 that the next scratch-it lottery ticket that the person buys is a winner, then the combined probability that the person has one brother and will win on the next scratch-it ticket is 0.3 × 0.2 = 0.06. The logic of the rule may become clear from the Venn diagram for the multiplication rule.

In the Venn diagram, A is the event a person has one brother and B is the event the person's next scratch-it ticket wins. We say that events A and B are independent if the chance of B being true, knowing that A is true, is just the same as the chance of B being true overall regardless of whether A is true. Clearly, knowing a person has one brother doesn't affect his or her chance of winning the lottery, so A and B here are independent. This is shown on the diagram by drawing the area of overlap (the probability of A and B occurring together) in the same ratio to the area of A as the area B is to the whole rectangle. The area A&B is 20 percent of the area of A, and the area of B is 20 percent of the area of the entire rectangle.

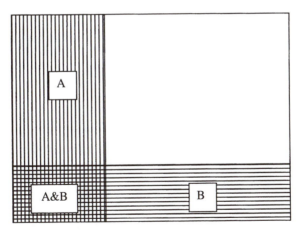

Put another way, since 0.2 of the time a person's next scratch-it ticket will win and this is true regardless of whether they have exactly one brother or not, and since their chance of having one brother is 0.3, on 0.2 of the occasions in which the 0.3 chance of having one brother eventuates, the person will also buy a winning scratch-it ticket. The chance of both events occurring is therefore 0.2 of 0.3, or $0.2 \times 0.3 = 0.06$, and this is the area represented by A&B on the Venn diagram.

The multiplication rule for independent events can also be extended to the case where there are more than two events. If we toss four fair coins once (or toss a single fair coin four times), whether we get a head on one coin cannot physically affect the outcomes on other coins, so tosses of coins are independent. On each toss the probability of a head is ½. By the extended multiplication rule for a series of independent events, the overall chance of four heads in a row is $\frac{1}{2} \times \frac{1}{2} \times \frac{1}{2} \times \frac{1}{2}$.

One way of convincing yourself of the sense of this rule is to consider sixteen tosses of four coins at a time. Say the coins are labelled A, B, C, and D. Out of the sixteen sets of coin throws we would expect there to be, on average, eight sets of throws in which coin A comes up heads. Out of these eight sets of throws we would similarly expect there to be, on average, four sets of throws in which coin B comes up heads. Out of these four sets of throws we would similarly expect there to be, on average, two sets of throws in which coin C comes up heads. Out of these two sets of throws we would similarly expect there to be, on average, one set of throws in which coin D comes up heads. Therefore, on average we would expect just one set of throws of the four coins out of sixteen sets of throws to deliver all heads. The probability of all heads is therefore 1/16, which matches the result $\frac{1}{2} \times \frac{1}{2} \times \frac{1}{2} \times \frac{1}{2}$ for the extended multiplication rule.

Often events are neither mutually exclusive nor independent. For example, knowing that a person is male decreases the chance of that person being short (compared to the overall average height of males and females combined), so the events "male" and "short" are not independent. On the other hand, the events "male" and "short" are not mutually exclusive, since it is possible for a person to be both "male" and "short." This situation will be discussed later.

Probability That an Event Does Not Occur: "not"

One other basic rule of probability that we will use is that the probability that an event does *not* occur is 1 minus the probability that the event does occur. If the probability of having exactly one brother is 0.4, the probability of not having exactly one brother is 0.6.

Venn Diagram Conventions

Traditionally, Venn diagrams are drawn with ellipses rather than drawn with rectangles.

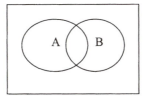

However, the use of rectangles rather than the traditional ellipses shows whether the area of overlap is in proportion to the area of both rectangles, as should be the case for independent events.

MORE COMPLICATED PROBABILITY RULES

The Modified Addition Rule

Often we want to find the probability of a combination of events that do not meet the criteria for the addition rule or the multiplication rule. The modified addition rule when events are not mutually exclusive is best explained by a Venn diagram.

Consider a lecturer in front of a large class closing her eyes and picking someone at random. Say we wanted to find the probability that the person picked will either be a woman or someone less than 1.6 m tall (or both). Take A to be the event that the chosen person is female and B to be the event that the chosen person is less than 1.6 m tall. These events are not mutually exclusive, since females less than 1.6 m exist and some such people presumably will be part of the class the lecturer is standing in front of.

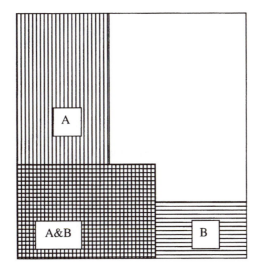

The diagram shows that if we added together the chance of choosing a woman (represented by area A, including the area A&B) to the chance of choosing a person less than 1.6 m tall (represented by area B, including the area A&B) we would be counting the chance of choosing a women less than 1.6 m tall (represented by area A&B) twice. To allow for this, the rule is "the probability of the event A or B or both occurring equals the probability that A occurs plus the probability that B occurs take away the probability that both occur together." The abbreviation for "or" in algebra is ∪ and the abbreviation for "and" is ∩ , so in symbols the rule is written as

$$P(A \cup B) = P(A) + P(B) - P(A \cap B)$$
$$(or) \qquad\qquad\qquad (and)$$

For example, if the probability of A (choosing a female) is 0.5, the probability of B (choosing someone less than 1.6 m tall) is 0.3, and the probability of A&B (choosing a woman less than 1.6 m tall) is 0.2, then the probability of choosing a person who is either a woman or who is less than 1.6 m tall or both is 0.5 + 0.3 − 0.2 = 0.6.

This is roughly represented by the Venn diagram. Here the area A (including the area A&B) is about 50 percent of the total area of the Venn diagram rectangle. The area B (including the area A&B) is about 30 percent of the total area of the rectangle. The area A&B itself is about 20 percent of the total area. We see that the sum of all the shaded areas representing the options of A or B or both is about 60 percent of the total area.

The Modified Multiplication Rule

The modified multiplication rule when events are not independent can be explained by reference to the same diagram. The modified multiplication rule gives us the probability of A and B occurring together if we know the probability of B occurring given the information that A has occurred (or vice versa). Take A to be the event a person is female and B to be the event that a person is less than 1.6 m tall. It makes sense to believe that the probability of finding a female person less than 1.6 m tall equals the probability of finding a person who is less than 1.6 m tall, knowing that the person is female, times the probability that person is female.

The abbreviation for "knowing that" (or "given that" or "given the information that") in algebra is an elongated vertical line, |, so that in symbols we write $P(A \cap B) = P(B|A) \times P(A)$. In a Venn diagram, $P(B|A)$ is represented by the proportion of area A that is also occupied by B. The jargon for $P(B|A)$ is "the conditional probability of B given that A is true." For example, say that 40 percent of females in the lecture are less than 1.6 m tall (i.e., the conditional probability of less than 1.6 m tall, given a female is chosen, is 0.4; this is represented on the diagram by the fact that the area A&B represents 40 percent of the total area of A, where A is taken to include the area A&B). Say also that the probability of choosing a female is 0.5. The probability that a person is chosen who is female and less than 1.6 m tall is then $0.4 \times 0.5 = 0.2$.

The Extended Modified Multiplication Rule

There is an extension to the modified multiplication rule to the situation of more than two events. Only one simple type of situation will concern us. Say we had to choose three balls blindly from a box that contained three red balls and one black ball. If each of the three choices were independent of the others, we would have a ¾ chance of choosing a red ball on each occasion, and by the simple rule of multiplication there would be a probability of $¾ \times ¾ \times ¾ = {}^{27}/_{64}$ of choosing all red balls. However, unless we replace each ball and shake the box after each choice, the three choices aren't independent: If we get a red on the first choice there are fewer reds to get on the later choices. To get three red balls in three choices, the first choice must be a red ball, but that leaves only two red balls to be chosen from three balls, a chance of $^2/_3$. The chance of two red balls out of the first two choices is then $¾ \times {}^2/_3$. So far this just illustrates the rule $P(A \cap B) = P(A) \times P(B|A)$. When it comes to the third choice we have only one red ball left out of two balls, so the probability of this choice giving a red ball as well is ½. The overall probability of all three choices giving red balls is then $¾ \times {}^2/_3 \times ½ = ¼$.

The same idea—the extended rule of multiplication—can be extended to the case of choosing a greater number of red balls from a box containing these red balls together with any number of black balls. If there are five red balls and nine balls in total in the box, the chance that we would get just the five red balls when we chose five balls at random from the box is $^5/_9 \times {}^4/_8 \times {}^3/_7 \times {}^2/_6 \times {}^1/_5$.

If there are k red balls and n balls total in the box, the chance that we would get just the k red balls when we chose k balls at random from the box is

$$\frac{k}{n} \times \frac{k-1}{n-1} \times \frac{k-2}{n-2} \times \frac{k-3}{n-3} \times \ldots \times \frac{k-(k-2)}{n-(k-2)} \times \frac{k-(k-1)}{n-(k-1)},$$

which can be rewritten as

$$\frac{k \times (k-1) \times (k-2) \times \ldots \times 3 \times 2 \times 1}{n \times (n-1) \times (n-2) \times \ldots \times (n-k+3) \times (n-k+2) \times (n-k+1)}.$$

APPLICATION OF THE EXTENDED MODIFIED MULTIPLICATION RULE TO FISHER'S EXACT TEST

We now have enough probability theory to explain our first statistical test. The test is Fisher's exact test. It applies when we have individuals that can be categorized by two separate methods and where each method of categorization divides individuals into just two categories. In this situation, we might want to know if it is reasonable to believe that the way an individual is categorized by one method is linked to the way the individual is categorized by the other method. To explain, consider an example.

	Fired	Not Fired	Totals
Women	3	0	3
Men	0	2	2
Totals	3	2	5

Say that a new male manager comes to a workplace and fires three of the five workers. Suppose that three of the original five workers were women and it turns out that these are the three who get fired. Here one method of categorizing an individual is as a woman or a man, and the other method of categorizing an individual is as a worker who is fired or as a worker who is not fired. This information is most conveniently displayed in a table. In this situation we may want to make a decision about whether we believe that being in the category "women" is linked to the category of "fired." Of course, the categories are linked in this particular instance. However, the manager, who is likely to be accused of sexism, may argue that this current instance should not be taken as a reflection of his general attitude. He may argue that, in general, he is not at all biased against women workers and that it was just by chance alone that it turned out that in this particular instance the workers who were fired were all women (we will assume that all the workers were equally competent and diligent and performing the same job, but economic circumstances dictated the firing of three of them). Fisher's exact test is designed to answer the question, "Should we believe the manager is sexist?"

Fisher's exact test, like other statistical tests, takes as its starting point the idea that until we are persuaded otherwise we should believe that there is no general linkage between the categorizations, and that the linkage apparent in this particular instance is just a chance event. In other words, in applying Fisher's exact test we start by believing idea 1: The manager ignored gender when he fired the workers and it was just chance that the fired workers were all women." Fisher's exact test then asks the question, "What is the size of this chance?" If it turns out that the chance is small, the sort of chance that "hardly ever" occurs, then statistics says that it is more reasonable not to blame the linkage in this particular instance on a chance that "hardly ever" occurs. Instead, it is more reasonable to believe idea 2: There is a real reason for the linkage. In other words, if it would hardly ever happen that all three women workers would be fired if three workers were chosen blindly from five, then it is reasonable to believe that the manager has a gender bias. There is a convention that says that the size of the chance that is so small that it "hardly ever" occurs is any chance that occurs less often than one time in twenty (in mathematical symbols, we write $p \leq 0.05$).

Our task is then to work out the chance that it would just be the three women workers who were fired if the three workers to be fired were chosen blindly from the five workers. This is exactly the chance that only the three red balls would be picked when choosing three balls blindly from a box containing three red balls and two black balls. Using the reasoning already explained, this chance is $^3/_5 \times {}^2/_4 \times {}^1/_3 = {}^1/_{10} = 0.1$. Since 0.1 is greater than 0.05, according to the conventions of statistics and the calculations of Fisher's exact test we would conclude that the manager's actions in this instance do not provide convincing evidence that he is sexist. Often a distorted form of words is used when a statistical test fails to provide convincing evidence, as in this situation. In this situation the distorted form of words might be "the figures show no evidence of sexism," or, even worse, "statistics show that the manager is just as likely to sack male workers as he is to sack female workers" or "statistics prove that there is no sexism involved."

Even the more carefully worded orthodox conclusion that there is no convincing evidence that the manager is sexist may not be reasonable. For a start, Fisher's exact test, like all statistical tests, started with the assumption that the manager should not be regarded as sexist unless convincing evidence was found. There is no good reason why this should be regarded as a reasonable starting point. There is also no good reason to use the value 0.05 as the size of the chance that "hardly ever" occurs. Many people would be reasonably convinced that there is a gender bias, given that the only other way we can account for the sackings of just the women workers is to argue that a 0.1 chance came off. The issues of an appropriate starting point and an appropriate p value arise because we want an answer to the main question: Is the manager sexist? But this is not the question we have answered. Instead, we have obtained a precise answer to a secondary question: If the manager was not sexist, how often would pure chance alone lead to the apparently sexist sackings that we observed? We have then leapt from a precise answer to this secondary question to a conclusion about the main question. In taking this leap we have not used any judgment or common sense appropriate to the situation. We have simply used an arbitrary convention that says that if the chance or p value in response to the secondary question is ≤ 0.05 we should state we have convincing (or "statistically significant" or "statistically significant at the 5% level") evidence that the answer to the first question is "yes." Otherwise we should state we don't have convincing evidence. Indeed, the phrase "convincing evidence" here is generally shortened to "evidence," so if $p > 0.05$ we state "we have no evidence." Later in this chapter we will determine the precise mathematical link between the main question, "Is the manager sexist?" and the secondary question, "Assuming the manager is not sexist, how easy is it for coincidence alone to explain the manager's action?" We will come to a precise conclusion about the manager. However, we first need to cover two more rules of probability: the law of total probability and Baye's theorem.

Before continuing, we will consider one practical objection that can be raised with our example. Although I stated that we assume that all the workers were equally competent and diligent, it must be admitted that unlike balls chosen from a box, individual workers are not identical. Firings of individual workers are not usually entirely at random. It will therefore be possible for the manager to point to some attributes of the workers he fired in order to justify his actions: Maybe they had less experience. This does not change the statistical analysis, but to the extent that we are willing to believe the justifications of the manager it will change how we word our conclusions. If we entirely believe his justifications, in place of the possible conclusion "the manager is sexist" we should write, "In workplaces supervised by this manager, women tend to be less valuable employees." The issue, then, is whether this example provides convincing evidence of this possibility. A way of phrasing a possible conclusion that is neutral between "the manager is sexist" and "in workplaces supervised by this manager, women tend to be less valuable employees" would be to state, "This manager tends to fire women." "The manager is sexist" is the formulation used in our example because it allows a concise form of words in a lengthy explanation and because it matches my own world view. While I am unrepentant about my choice, there is a lesson here: It is easy to load a statistical analysis with ideological bias.

Now consider the scenario discussed in Chapter 1 of music students and one-half or one hour tuition per week. Let us say the outcome we observed was not the students' actual marks on their music exam, but simply whether they passed or failed. Say the outcome is as given in the accompanying table.

	Failed	Passed	Totals
½ hour lesson	3	0	3
1 hour lesson	0	2	2
Totals	3	2	5

As in the previous example, there are two possible explanations for the observed outcome: (1) The longer lessons are of no more use than shorter lessons in helping students pass (it just looks that way purely as a result of blind chance) or (2) longer lessons do help students pass.

Again, as in the previous case, we start by assuming option 1 is the correct explanation and using this assumption find out the size of the chance involved. The calculation is exactly the same as for the firing of the women, so, as before, $p = 0.1$. Again, since $p > 0.05$, frequentist statistics would want us to state, "There is no convincing evidence that one-half-hour lessons are less effective than one-hour lessons." This is a strange conclusion, even less appropriate than the conclusion that statistics tells us we should draw in the case of the manager and the firing of the women. This is because the starting point that extra music tuition is of no benefit is even more dubious than the starting point that the manager is not sexist. Indeed, common sense tells us that the starting point that extra music tuition is of no benefit is almost certainly wrong. Of course, it is possible to argue that at some point forcing a child to spend more time in lessons will be counterproductive and that this point is reached at just the right place between half an hour and one hour that the negativity caused by the longer lesson exactly balances its positive value. This seems very unlikely. Common sense suggests that there is likely to be an advantage in having one hour rather than half an hour of tuition per week. Therefore, although if we wanted to we could explain the results by arguing that extra tuition makes no difference, it just looks that way because a 0.1 chance came off; there is no point in putting this argument.

Strangely, though, many people who use statistics are unaware of the frequent conflict between common sense and conclusions based solely on the $p \leq 0.05$ tradition of frequentist statistics. Even many research scientists do not have an appreciation of the need to use common sense with statistics. As a result, the scientific literature is full of inappropriate conclusions: For example, "There is no relationship between domestic violence and social class," "Lowering the speed limit by 5 mph has not changed the accident rate," "Greater attention to hygiene makes no difference to the chance of infection," "Students in large classes do just as well as students in small classes." The list is endless. In most cases the data would have shown the expected relationship between the cause and the likely effect, despite the bland denials used in the conclusions. The misleading bland denials are inappropriately justified by the fact that if one wanted to one could explain the relationships seen in the data by arguing that a chance with probability > 0.05 came off, which just happened to give the appearance of the expected relationship. If the actual relationships are very strong and the amount of data is very large, it is most unlikely that by sheer bad luck we would obtain such weak data that the relationship seen in the data could be explained away by appeal to a chance occurring with probability > 0.05. Therefore, statistics will lead us to appropriate conclusions in such circumstances.

However, collecting data is expensive. Often the amount of data obtained will not be very large and there is no general rule about the amount of data that must be examined before a researcher is able to conclude that there is no relationship. Many of the examples I use involve even less data than would generally be used in research. These examples are used not only to keep calculation simple, but also to emphasize how inappropriate it may be to declare that there is no evidence when $p > 0.05$. The issue of the amount of data that should be obtained is discussed further in Chapter 6 and Chapter 10.

Let us consider another example where Fisher's exact test could be used. Say four men and five women are interviewed about butter versus margarine preference and all the men prefer butter and all the women prefer margarine.

	Male	Female	Totals
Prefer butter	4	0	4
Prefer marg	0	5	5
Totals	4	5	9

Does this provide convincing evidence of gender bias in butter–margarine preferences? In this example no controversial issues are involved. The starting point—that there is no gender preference—seems reasonable, though on the other hand, since women may be more concerned with nurturing and possibly therefore with health issues, it is possible that women more than men will tend to avoid butter. We will simply deal with the calculations of the test.

The philosophical shortcomings of the statistical test are not such an overwhelming issue here, as the starting point seems reasonable. Again, we calculate an answer to the secondary question. That is, we assume for the time being that there is no particular tendency for men to be butter preferrers, and using this assumption we calculate the probability that pure chance alone would lead to the observed result. Here we could regard the four men as four red balls in the box and the four butter preferrers as the four balls we are choosing from the box. The chance that just the red balls (the men) would be chosen (to be butter preferrers) is then $p = \frac{4}{9} \times \frac{3}{8} \times \frac{2}{7} \times \frac{1}{6} \approx 0.008 < 0.05$. (Regarding the five women as five red balls and the five who are margarine preferrers as five choices from the box also gives $p = \frac{5}{9} \times \frac{4}{8} \times \frac{3}{7} \times \frac{2}{6} \times \frac{1}{5} \approx 0.008$.) Since both methods of calculating the p value are valid they must give the same answer. Since $p \leq 0.05$, statistics would tell us we should conclude that the figures give convincing evidence that gender influences butter–margarine preferences.

In the examples so far the calculations have been easy. However, the calculations would be much more difficult if our survey of gender and butter–margarine gave us the result in the accompanying table.

	Male	Female	Totals
Prefer butter	10	5	15
Prefer marg	10	15	25
Totals	20	20	40

The situation is more complicated here because we don't have zeros in the diagonal cells. Here, half the men but only a quarter of the women prefer butter. How do we calculate the answer to the secondary question now? In other words, how do we calculate how often this result would arise when only pure chance, not gender preference, is involved? The complications that arise when we don't have zeros on the diagonals are dealt with in Chapter 8, but the basic philosophical principle behind the calculation is unchanged. For those who have already come across various statistical tests, we note that another test called the chi-square or χ^2 test is commonly used in situations similar to those in which we have used Fisher's exact test. Again, the philosophical principles involved in interpreting the p value are unchanged, but there are slightly different assumptions involved in the calculation.

APPLICATION OF THE EXTENDED MODIFIED MULTIPLICATION RULE TO COMBINATIONS

Let's return to the question of the chance of choosing k red balls from a box containing k red balls and a total of n balls. As well as being relevant to Fisher's exact test, the problem is relevant to other statistical tests. We already have a mathematical formula for the probability of choosing just the k red balls: From p. 39 it is

$$\frac{k \times (k-1) \times (k-2) \times \ldots \times 3 \times 2 \times 1}{n \times (n-1) \times (n-2) \times \ldots \times (n-k+3) \times (n-k+2) \times (n-k+1)}.$$

Another answer is that there is only one particular set of k balls that are all red, but there are a number of other sets of k balls where not all the balls are red. If k balls are chosen blindly, all possible sets of k balls should be equally likely to occur. The chance of choosing the k red balls could then be stated as one way out of all the ways there are of choosing k balls from n balls. The number of ways there are of choosing k balls from n balls is given the symbol ^nC_k (the C is referred to as the combinations symbol; ^nC_k can be pronounced as "n Choose k"). The chance of choosing the k red balls is then $1/^nC_k$. Equating this chance with the chance given by the mathematical formula, we have

$$\frac{1}{{}^nC_k} = \frac{k \times (k-1) \times (k-2) \times \ldots \times 3 \times 2 \times 1}{n \times (n-1) \times (n-2) \times \ldots \times (n-k+2) \times (n-k+1)}.$$

Turning the fractions on both sides of the equality upside down gives us the formula for the number of ways of choosing k objects from a collection of n objects, ${}^nC_k$:

$$ {}^nC_k = \frac{n \times (n-1) \times (n-2) \times \ldots \times (n-k+2) \times (n-k+1)}{k \times (k-1) \times (k-2) \times \ldots \times 2 \times 1}.$$

Another method of deriving the formula for ${}^nC_k$ is given at the end of this chapter. (Here we use the logic that says that if $1/x = 1/2$ then $x = 2$.)

There is an appreciable branch of mathematics dealing with various complexities and relationships satisfied by expressions involving the combinations symbol. We will give just one: ${}^nC_k = {}^nC_{n-k}$. A one-sentence logical proof follows. ${}^nC_k$ is the number of ways for a teacher to choose k children from a class of n to stand up; logic dictates that this must be the same as the number of ways of choosing $n-k$ children to remain seated (${}^nC_{n-k}$) with the remainder of the class (k children) to stand up. For example, there are ${}^{25}C_5$ ways of choosing five children in the class of twenty-five to stand up with the rest to be seated, and there are ${}^{25}C_{20}$ ways of choosing twenty to sit down with the remaining five to be standing. Logic dictates that since in both cases the result is all the ways of having five children standing and twenty sitting down, the number of ways this can be done must be the same regardless of whether we call it ${}^{25}C_5$ or ${}^{25}C_{20}$.

OPTIONAL ▨▨▨

Here is another example of the use of Fisher's exact test with a solution using the combinations symbol. Imagine there is a survey of n areas in a national park; k of them are near a walking track and the others are not. Say that introduced weeds were found only in the k areas near the walking track. Someone might want to argue that this finding is pure coincidence. There are again two possible choices: (1) It is coincidence, in which case the finding must be attributed to a $1/{}^nC_k$ chance coming off (there are ${}^nC_k$ possible choices of k areas out of the n surveyed, but just one of those choices consists solely of the k areas near the walking track); or (2) the relationship between weeds and nearby paths is not due to chance alone. The decision as to which is the most reasonable explanation is up to us, but it needs to be guided by the evaluation of the chance $1/{}^nC_k$. Statistical tradition tells us to believe that option 2 is correct if $1/{}^nC_k$ turns out to be ≤ 0.05, and otherwise tells us to state that there is no convincing evidence that option 2 is correct, but this tradition does not necessarily accord with common sense, so it is important we make up our own mind. Say $n = 8$ and $k = 3$; then ${}^nC_k = 56$, and so $1/{}^nC_k \leq 0.05$. In this case most people using common sense would agree with someone making the decision solely on the basis of statistical convention. It seems almost certain, given these figures and using common sense, that weeds in an area have something to do with proximity to walking tracks.

As well as using the formula for ^nC_k, we can use a direct argument to calculate the chance $1/^nC_k$. We start with the assumption that the distribution of weeds is due to random chance, but we are aware that three out of eight areas are infested. Then when we look at the first area close to the track we would say that it is weed infested because a $^3/_8$ chance came off (three weed areas out of eight areas in total), when we look at the second area close to the track we would say that it is weed infested because a $^2/_7$ chance came off (two remaining weed areas out of seven remaining areas), and when we look at the third area close to the track we would say that it is weed infested because a $^1/_6$ chance came off (one remaining weed area out of six remaining areas). The overall chance is found by multiplying these probabilities: $^3/_8 \times ^2/_7 \times ^1/_6 = ^1/_{56}$ (formally, we are multiplying out conditional probabilities to get the combined probability).

END OPTIONAL

THE LAW OF TOTAL PROBABILITY

Often we have a population subdivided into groups and we know the chance of something happening (or the proportion of the time something happens) in each group. The law of total probability then gives us the overall chance of that thing happening (or the overall proportion of the time that something happens).

An important example concerns medical tests. The most accurate medical tests often involve direct laboratory examination of a portion of the organ that is affected by a disease. These tests are usually invasive; in other words, they involve operations and so are expensive and somewhat dangerous. However, in most cases there are preliminary tests, such as blood or urine tests, based on effects the diseased organ has on ingredients in the blood or urine. These less direct indicators of disease are often less reliable, though often these tests are worthwhile because the disadvantage of reduced accuracy is outweighed by the benefits of the avoidance of an operation.

The preliminary tests can be unreliable in two ways. They can give the wrong answer when the person has the disease and they can give the wrong answer when the person doesn't have the disease. There are two terms relevant here. The sensitivity of a test is the proportion of the time that a test gives the correct answer when the person has the disease. The specificity of a test is the proportion of the time that a test gives the correct answer when the person doesn't have the disease. Since all positive preliminary tests (i.e., tests indicating disease) have to be followed up with more invasive and expensive tests, it is often of interest to know how many positive preliminary tests will arise in any testing program. To do the calculation we have to know the sensitivity and specificity of the test and the proportion of the population that has the disease. Relating this back to the introductory paragraph defining the law of total probability, here the groups in the population are those with the disease and those without, and the sensitivity and specificity let us know the proportion of time something happens (a positive test) in each group.

Say the specificity of a test is 0.95. This value for specificity is often used in blood tests where the concentration of some ingredient varies from person to person. By tradition, those normal people who are unusual enough to have concentrations in the highest 5 percent are often considered to need further testing to differentiate them from diseased people who also have high levels. For brevity, we will summarize this statement by $P(T-|ND) = 0.95$ and read this as the probability that the *Test* is "–" or negative given *No Disease* = 0.95. Since 95 percent of those with no disease test negative, 5 percent test positive, so we can write $P(T+|ND) = 0.05$. Say also that the sensitivity is 0.8. Then, using similar notation and reasoning, we write $P(T+|D) = 0.8$ and $P(T-|D) = 0.2$. Let us also say that we are testing a population where 10 percent have the disease. We write this as $P(D) = 0.1$ and $P(ND) = 0.9$. The question, then, is to find out the proportion of positive tests that result when we test this population.

The answer can be obtained using common sense aided by a diagram (see Figure 3.1). The proportion who test positive consists of those who don't have the disease but test positive and those who do have the disease and test positive. These proportions are 5 percent of 90 percent and 80 percent of 10 percent, or 0.045 and 0.08, respectively. The overall proportion who test positive is then 0.045 + 0.08 = 0.125, or 12.5 percent.

The law of total probability here can be stated as follows: The overall probability of a positive test equals the probability of a positive test, knowing that there is no disease times the chance that there is no disease plus the probability of a positive test knowing there is disease times the chance that there is disease. In symbols, the law of total probability here can be written as follows:

$$P(T+) = P(T+|ND) \times P(ND) + P(T+|D) \times P(D)$$
$$0.125 = \quad 0.05 \quad \times \quad 0.9 \quad + \quad 0.8 \quad \times 0.1$$

The law of total probability can be easily generalized to situations where there are more than two groups in the population, but we will not be concerned with such situations.

OPTIONAL

As a second example of the use of the law of total probability, say a plane disappears and it is thought that there is a 1-percent chance that it has crashed in the sea. If so, there is only a 2-percent chance that it will be found. However, if it has crashed on land there is a 90-percent chance that it will be found. What is the overall probability that it will be found?

Figure 3.1

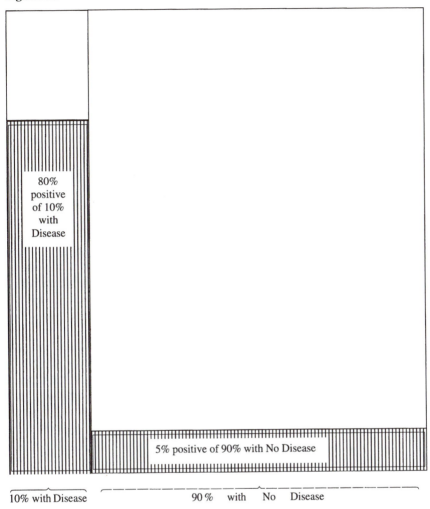

80%
positive
of 10%
with
Disease

5% positive of 90% with No Disease

10% with Disease   90 %   with   No   Disease

In answering this question it is helpful to use obvious notation, such as P(F|S) = 0.02 for the 2-percent chance that it will be Found given that it has crashed in the Sea. With such notation, the law of total probability gives us P(F) = P(F|S) × P(S) + P(F|L) × P(L), so the probability that it will be found is 0.02 × 0.01 + 0.9 × 0.99 = 0.8912; that is, there is a 89.12-percent chance that it will be found.

END OPTIONAL

BAYES'S RULE

In the first instance, Bayes's rule is just a way of swapping around the events in a conditional probability statement. Say we know the sensitivity of a medical test—we know the probability of a positive test for a disease (T+) given that a person has the disease (D)—but the problem of interest is the probability that a person has the disease given that the test is positive; that is, we have $P(T+|D)$ but we want $P(D|T+)$. From p. 38 we know that $P(T+ \cap D) = P(T+|D)P(D)$ (recall that the symbol $\cap$ means "and," so in words we are stating that the probability of a person having a positive test and having the disease equals the probability of a person having a positive test knowing that the person has the disease times the probability of the person having the disease). We also know, by symmetry, $P(T+ \cap D) = P(D|T+)P(T+)$ (in words, the probability of a person having a positive test and having the disease equals the probability of a person having the disease knowing that the person has a positive test times the probability of the person having a positive test), so both $P(T+|D)P(D)$ and $P(D|T+)P(T+)$ equal $P(T+ \cap D)$. We therefore have

$$P(D|T+)P(T+) = P(T+|D)P(D)$$

so

$$P(D|T+) = \frac{P(T+|D)P(D)}{P(T+)}$$

Often the $P(T+)$ in the denominator is expressed in terms of the law of total probability: $P(T+) = P(T+|ND) \times P(ND) + P(T+|D) \times P(D)$. Using this last equation gives us Bayes's rule as it would commonly be written in this situation:

$$P(D|T+) = \frac{P(T+|D)P(D)}{P(T+|D)P(D) + P(T+|ND)P(ND)}$$

For example, consider a medical test with a sensitivity of 0.8 and specificity of 0.95 applied to a population in which 10 percent have the disease. What is the probability that a person has the disease given that they test positive? The answer is

$$P(D|T+) = \frac{0.8 \times 0.1}{0.8 \times 0.1 + 0.05 \times 0.9} = 0.64$$

or about $^2/_3$. Referring to Figure 3.1, what these calculations are doing is stating that we know that we are dealing with a positive test so we know that diagrammatically we are in the shaded area. Knowing that we are within this shaded area, we want to know the chance that we are in the left-hand column constituting 10 percent of the Venn diagram and corresponding to those who have the disease. The calculations tell us that about $^2/_3$ of the shaded area is in the left hand column corresponding to the disease region.

Let us see what would happen if only 1 percent of the population had the disease (see Figure 3.2). Then $P(D|T+)$ would be or approximately 0.14 or about one-seventh. In other words, only about one-seventh of the people who tested positive for the disease would actually have the disease. The test is surprisingly inaccurate in this situation because the small proportion of false

Figure 3.2

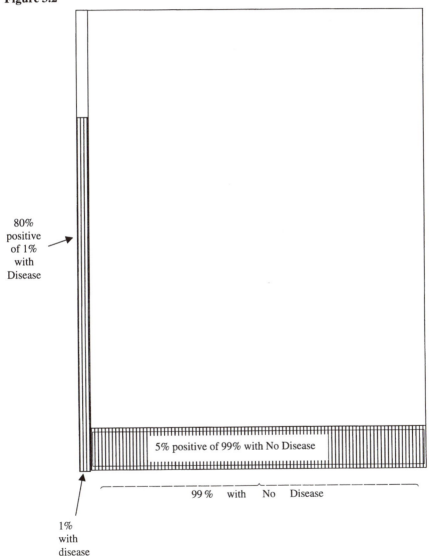

80% positive of 1% with Disease

5% positive of 99% with No Disease

99% with No Disease

1% with disease

positives in the disease-free population who constitute the vast majority outweighs the true positives in the tiny minority who do have the disease. In fact, they outweigh this tiny minority by about six to one.

The issues raised in this medical example are of considerable practical importance. Screening tests for all but the most common of diseases may often be counterproductive because the costs of the initial screening plus the extra costs and dangers of the more invasive follow-up testing necessary for those who are "false positives" (T+|ND), may well outweigh the benefits of detecting the true positives (T+|D).

OPTIONAL

As another example, consider the example of the missing aircraft in the section on the law of total probability. What is the probability that it crashed on land given that it was not found? Using obvious notation, we have $P(\text{NF}|S) = 0.98$, $P(\text{NF}|L) = 0.1$ and $P(L) = 0.99$. Then

$$P(L|NF) = \frac{P(NF|L)\,P(L)}{P(NF|L)\,P(L) + P(NF|S)\,P(S)} = \frac{0.1 \times 0.99}{0.1 \times 0.99 + 0.98 \times 0.01} = 0.9099$$

That is, whereas initially we had the idea that the a priori chance was 99 percent that it had crashed on land, now, with the knowledge that it was not found, there is about a 91 percent a posteriori chance that it is on land.

END OPTIONAL

The same sort of reasoning can be applied in many other situations. In law we might know the chance of some piece of evidence (e.g., a particular blood group) being associated with someone who is innocent and the chance of that particular evidence being associated with the guilty person. If we also have some prior estimate of the chance of guilt of the person before the court, then knowledge of the presence of the evidence can be used to improve our estimate of guilt in a mathematically precise way. Unfortunately, despite several hundred years of effort by probability theorists, this quantitative approach to assessing guilt has not generally been taken on by the legal profession.

BAYES'S RULE AND THE ASSESSMENT OF THE RESULTS OF STATISTICAL TESTS

For our purposes, the most important use of Bayes's rule relates to its use in incorporating common sense into assessment of the results of statistical tests. Let us consider the earlier example of Fisher's exact test in the case of the manager who fired all three women in a group of five workers. As we have seen, although we may want to know whether the manager is sexist, this is not the question directly answered by statistics. Instead, statistics tells us that if the manager was not sexist, we could expect that such a result would happen with probability 0.1. In other words it would happen by chance alone 10 percent of the time. We then use the convention in statistics that 0.1 is more often than the "hardly ever" chance of 0.05. The convention then tells us to take a leap, so we state that the answer to the original question, "Is the manager sexist?" is that "there is no (convincing) evidence." Let us see how Bayes's rule could allow us to give a more satisfactory answer.

We will assume that if the manager is sexist there is a 100-percent chance that if he is required to fire three workers he will choose the women workers. We will also assume that we believe that 40 percent of managers are sexist. In obvious notation, what we have is $P(\text{FW}|S) = 1.0$ (probability of *Firing Women* given *Sexist* manager = 1). From Fisher's exact test we have $P(\text{FW}|NS) = 0.1$ (probability of *Firing Women* given *Non-Sexist* manager = 0.1). We also have, from our belief, that $P(S) = 0.4$ (probability of *Sexist* manager = 0.4), so $P(NS) = 0.6$ (probability of *Non-Sexist* manager = 0.6). We can now use Bayes's rule to calculate an answer to the question of interest: Do the figures indicate the manager is sexist? More precisely, we calculate $P(S|\text{FW})$ (probability of *Sexist* manager given *Fired Women*). Bayes's rule gives us

$$P(S|FW) = \frac{P(FW|S)\,P(S)}{P(FW|S)\,P(S) + P(FW|NS)\,P(NS)} = \frac{1 \times 0.4}{1 \times 0.4 + 0.1 \times 0.6} \approx 0.87.$$

In other words, we now have a precise answer. There is an 87-percent chance that the manager is sexist.

The obvious question here is why not use Bayes's rule all the time, seeing that it is such a sensible and precise refinement in decision making using statistics. We will discuss this issue in the context of the example of the women firings and the manager, though the same ideas apply generally. The answer is that the use of Bayes's rule involves some subjective judgments, whereas no subjective judgments are needed in the calculation that $P(\text{FW}|NS) = 0.1$. The quest for objectivity led early statisticians to avoid using Bayes's rule explicitly. Instead, it seemed more reasonable to work out objective values like $P(\text{FW}|NS)$ and leave readers of the scientific papers to do a rough Bayes's rule calculation in their head using their own ideas about values like $P(\text{FW}|S)$ and $P(S)$. Unfortunately, this turned out to be an unreasonable expectation of the readership of scientific papers. The rather awkward mental assessment of the implication for $P(S|\text{FW})$ (probability *Sexist* given *Fired Women*) of a particular value of $P(\text{FW}|NS)$ (probability *Fired Women* given *Not Sexist*), then became stylized. It became the traditional statistics formula that if $P(\text{FW}|NS) \leqslant 0.05$, then believe S (the manager is sexist) given the event FW (*Fired Women*). If $P(\text{FW}|NS) > 0.05$, don't believe S given the event FW (or at least state there is no convincing evidence for it). In defense of this frequentist statistics approach, it should be pointed out that our prior subjective ideas that

$P(S) = 0.4$ and $P(FW|S) = 1$ are entirely personal. We have only personal belief to specify that $P(S) = 0.4$. It may also not be reasonable to assume that $P(FW|S) = 1$, since a sexist manager may avoid drawing attention to himself by firing only two of the women. Different values inserted by different people will result in different values for $P(S|FW)$.

In many instances calculations using Bayes's rule can also be much more difficult. Although, many statisticians favor the use of Bayes's rule in statistics (Bayesian statistics) or some related approach, frequentist statistics that stops at calculating $P(FW|NS)$ is still by far the main approach to statistics. As we have seen, this frequentist statistics, if it is coupled firmly to a rule like "convincing evidence of S given FW if $P(FW|NS) \leq 0.05$, otherwise "no," often leads to inappropriate conclusions.

However, if we are prepared to uncouple ourselves from this rule and use common sense to decide the value of $P(FW|NS)$, at which we become convinced of S given FW, then we can use frequentist statistics wisely. If we assess the value of $P(FW|NS)$ in this way we are in effect using a mental calculation of Bayes's rule. This does not mean that we literally perform, as mental arithmetic, the calculations involved in Bayes's rule. Instead, what is meant is that we should always ask ourselves after a statistical test which idea is more reasonable to believe. In the case of the fired women, we should ask ourselves, "Is it more reasonable to believe that the manger is not sexist but it just looks that way because a one in ten chance came off when he made a decision as to which workers should be fired oblivious to gender, or, is it more reasonable to believe that the manager is sexist?" In other words, to use statistics sensibly, in each situation we should use common sense to decide the p values that for us would amount to convincing evidence.

In the past this may have been thought to be too much to ask of the readership of scientific journals. However, the widespread inappropriate use of statistics in many areas of human activity means that it is important that people learn to incorporate common sense into statistics. There has been some limited recognition of the need for this approach. In the past, scientific papers often just stated that results were "statistically significant," meaning p was $\leq$ 0.05. Now most articles in scientific journals quote p values and so potentially allow the readership to make up their own mind about "statistical significance."

We will continue to emphasize the importance of incorporating common sense into statistical decisions. However, in most situations a complete application of Bayes's rule to analysis of statistical tests is difficult. We will therefore not explicitly apply the principles of Bayesian statistics to other statistical tests. Instead, we will emphasize the need to make our own common-sense value judgments in each situation, about the p values that we can fairly regard as convincing evidence.

In the previous examples of Fisher's exact test we have sometimes disagreed with the conclusions that would have been reached using the traditional p value benchmark of 0.05 because we have been dealing with situations where it seemed appropriate to be more easily convinced. For example, in the case of the manager who fired just his three women workers, my guess is that most people would think that the man was sexist, particularly after working out that a manager who ignored gender would have only one chance in ten of sacking only the women (i.e., $p = 0.1$). Here, even though $p = 0.1$ (> 0.05), it would still be seen in this context as reasonably convincing evidence.

We now come to an example where most people would not be convinced, even though the p value is much smaller than 0.05. Say the next-door neighbor's child claimed to be a clairvoyant. She produces a box containing sixteen identically shaped balls, three of which are white and the rest red. In particular, she claims to be able to pull just the white balls out of the box blindfolded. Without looking, she chooses three of the balls. In fact, it turns out that she does get just the white balls. Should we believe that she is a clairvoyant?

	Red	White	Totals
Chosen	0	3	3
Not Chosen	13	0	13
Totals	13	3	16

There are only two options: (1) result is due to random chance, or (2) result is due to clairvoyance (we assume that the possibility of a magic trick has been excluded and the difference between red and white balls cannot be detected by touch). First we work out the size of the chance, assuming that option 1 is true. Using the same probability rules as before, we see that the probability of this happening is $1/^{16}C_3 = 1/560$. If we follow the tradition of statistics, since $1/560 \leq 0.05$ we would believe that the child is a clairvoyant. We should, however, use our common sense to decide which of the two options is more likely. My own opinion would be that option 1 is far more plausible than option 2. It is my nature to be quite skeptical of phenomena such as clairvoyance in general, but even those who consult fortune tellers are likely to be dubious about the abilities in this regard of the next-door neighbor's child. Here, although $p \approx 0.002$ (< 0.05), most of us would disagree with the traditions of statistics and not regard this single demonstration as convincing evidence.

Let us apply Bayes's rule explicitly to this situation. Say that prior to seeing the child's performance we believed that the probability that the child had the ability of *C*lairvoyance was one in a million, so $P(C) = 0.000001$, and the probability of *N*ot *C*lairvoyance was one in 999,999, so $P(NC) = 0.999999$. AW denotes the event *A*ll *W*hite balls chosen. We have $P(AW|NC) = 1/560$, and we assume that $P(AW|C) = 1$. We can then use Bayes's rule to work out

$$P(C|AW) = \frac{P(AW|C)\,P(C)}{P(AW|C)\,P(C) + P(AW|NC)\,P(NC)} \approx 0.00056$$

That is, we conclude that there is a 0.056-percent chance that the child is a clairvoyant rather than just lucky. If the child got just the white balls again on a second go, a repeat application of Bayes's rule would give her a 24-percent chance of being a clairvoyant. A third successful go by the child at choosing only white balls and another repeat application of Bayes's rule would mean that I should now believe that there is a 99.4-percent chance that the child is a clairvoyant, although I would tend to suspect a magic trick.

AN ALTERNATIVE APPROACH TO COMBINATIONS

Number of Possible Arrangements of *n* Objects

Often probability calculations rely on the idea that it is sometimes reasonable to expect that all arrangements are equally likely to occur. For example, if ten names are put in a hat, what is the probability that they will be withdrawn in alphabetical order? There are ten ways of choosing the first, and for each of those ten ways there are nine ways of choosing the second, so there are 10×9 ways of choosing the first two. For each of these ninety ways of choosing the first two there are eight possible choices remaining for the third, so there are $10 \times 9 \times 8$ ways of choosing the first three, and so on. We discover that there are $10 \times 9 \times 8 \times 7 \times 6 \times 5 \times 4 \times 3 \times 2 \times 1$ ways of choosing all ten names. This number is denoted 10!, pronounced "ten factorial" (it is 3,628,800). Only one of these ways is alphabetical order, so the chance of alphabetical order is 1 in 3,628,800.

As another example, say five countries in a region are ranked according to their population growth rates and according to their amounts of poverty. It is found that the two rankings agree completely: Is it reasonable to believe that this agreement is due entirely to chance? To answer, say we were to blame the apparent correlation entirely on chance. Then the poorest country would have a choice of any of the five ranks for population growth, the next poorest country would have a choice of any of the four remaining ranks, and so on, so that the number of possible orderings would be $5 \times 4 \times \ldots \times 1 = 5! = 120$. There is only one of these 120 arrangements where the two rankings agree completely. It is then up to us to decide which is a more reasonable explanation for the agreement between the rankings. The choices are (1) that there is no underlying reason to explain the connection between the two rankings other than a $1/120$ chance coming off, or (2) that there is some underlying reason that explains why poorer countries tend to be associated with more population growth.

We still have to make a somewhat arbitrary decision between the two choices. However, the decision is guided by our calculation of the probability $1/120$ associated with choice 1. This is another example of a statistical test (in technical terms we are testing the null hypothesis that Spearman's rho is 0, but we won't bother with this jargon yet). It should be noted that if (as is reasonable) we decide to believe that there is an association between poverty and population growth that cannot fairly be attributed to chance alone, then we still have not proven that population growth causes poverty. It may be that poverty causes population growth, or some other component of the socioeconomic environment tends to cause poverty and also cause population growth. Probably all three factors operate to some degree in this situation.

Permutations

Sometimes it is necessary to know how many ways there are of choosing k objects in a particular order from a group of n objects. This is denoted ^nP_k, and using the same principles as earlier equals

$$n \times (n-1) \times (n-2) \times \ldots \times (n-k+1) = \frac{n!}{(n-k)!}.$$

This is $n!$ with the last $n-k$ factors cancelled out by the division. For example, say a teacher has a class of twenty students and needs to choose three to be the class captain, the blackboard monitor, and the representative on the student council. There are $20 \times 19 \times 18 = {}^{20}P_3$ ways of choosing a set of three children for these different jobs. We see that this can be written as:

$$\frac{20 \times 19 \times 18 \times 17 \times 16 \times 15 \times \ldots \times 3 \times 2 \times 1}{17 \times 16 \times 15 \times \ldots \times 3 \times 2 \times 1} = \frac{20!}{17!}$$

There are twenty choices for captain and for each of these twenty choices there are nineteen choices for blackboard monitor and then for each of these 20×19 combinations of choices there are eighteen possible students who could be chosen as student representative. Note that the situation makes order important. The first choice, the second choice, and the third choice are not equivalent. They all get different jobs.

Combinations

Often it is necessary to know how many ways there are of choosing k objects from a group of n objects when order is irrelevant. For example, three students are to be chosen for a committee, but they are not going to get different jobs as captain, blackboard monitor, and representative just because their names are selected in a particular order. Drawing out the names Sue, Mary, and John in that order from a hat containing all the names is counted as the same as drawing out in order the names John, Sue, and Mary. As stated previously, the number of ways of choosing k objects from n objects is denoted ^nC_k.

It can be calculated by reasoning that for each of the ^nC_k choices of k objects, $k!$ orderings are possible for these objects. This then gives all possible choices of k objects in order, which is ^nP_k. This tells us that $^nC_k \times k! = {}^nP_k$, so

$$^nC_k = \frac{^nP_k}{k!} = \frac{n!}{k!(n-k)!}$$

For example, say the teacher in the permutation example thought about making her choice of students for the three special jobs in the class (captain, blackboard monitor, and student rep) in the following way: She thought to herself, "How many ways can I choose three students from twenty? Whatever number this is, I'll call it $^{20}C_3$. Then, for each of these choices I'll bring the three students up to the front of the class. The three students can be assigned to the three jobs in 3! ways (three choices for captain and for each of these choices, two choices for blackboard monitor, and then just one choice for student rep). All up then there will be $^{20}C_3 \times 3!$ ways of assigning the three special jobs to the students." But this $^{20}C_3 \times 3!$ is counting the same number of arrangements as $^{20}P_3$. Therefore, we have found out indirectly that

$$^{20}C_3 = \frac{^{20}P_3}{3!} = \frac{20!}{17! \times 3!}$$

There is just one way of choosing all n children from a group of n children. Our notation tells us this number of ways is ^nC_n. Therefore we should have $^nC_n = 1$. The formula tells us that

$$^nC_n = \frac{n!}{n! \times (n-n)!} = \frac{n!}{n! \times 0!}$$

Mathematicians define 0! to be 1. The formula

$$\frac{n!}{n! \times 0!}$$

then makes sense and gives us the required result of 1.

SUMMARY

Probabilities combine according to rules whose logic can be seen from Venn diagrams. The logical rules can be combined to give more complicated probability rules. Two are of particular interest:

• Fisher's exact test uses probability rules to work out the probabilities of certain outcomes when individuals in two groups are assigned at random to one of two possible categories. If those in one particular group tend to be assigned to one particular category and probability rules shows that this outcome has a very low probability when individuals are assigned at random, it suggests that random assignment is not the explanation. However, common sense must also be used in each case in judging whether random chance is still the best explanation.

- Bayes's rule uses probability rules to combine prior guesses about alternative explanations, and probability calculations about how easily the different alternative explanations can result in the observed outcome, to give an overall probability of which explanation is correct. In theory, Bayes's rule can be used to combine p value calculations and common sense for better decision making.

Use Fisher's exact test to obtain a p value when dealing with a situation of two groups and two categories. This is a situation that can be summarized in the form of the accompanying table.

	Group I	Group II
Category A		
Category B		

Always use common sense as well as p values in coming to a conclusion about whether the groups and categories are associated.

QUESTIONS

1. There are eighty women and sixty men in the class. Thirty women and thirty men own dogs. The lecturer shuts his or her eyes, spins around, and points to a person at random.

   a. What is the probability that the person picked:

      i. Is a woman?

      ii. Owns a dog?

      iii. Is a female dog owner?

      iv. Is either a woman or a dog owner?

   b. Relate your answer in part a to the formula $P(E \cup F) = P(E) + P(F) - P(E \cap F)$.

   c. In this situation are the events "being a woman" and "being a dog owner" independent, mutually exclusive, or neither?

2. Human blood can be grouped in several different ways. The rhesus blood groups (D+ and D–) and the ABO blood groups are inherited independently of each other. About 14 percent of the population has the rhesus blood group D–, the remaining having D+, and 3 percent of the population has ABO blood group AB. Inheritance of AB blood group and O blood group is mutually exclusive; 55 percent of the population has blood group O.

    a. What proportion of the population has both AB and D–?

    b. What proportion has blood group AB or O?

    c. What proportion has blood group AB and/or blood group D–?

    d. What proportion has blood group O and/or blood group D+?

3. According to a news report, one-half of all women and one-third of all men over the age of seventy suffer from osteoporosis (unduly weak bones). Men make up one-quarter of this age group. Find the overall prevalence of osteoporosis in people over seventy (take "one-half" to mean a proportion of exactly 0.5, etc.).

4. Assume that you are told that 2 percent of children under the age of twelve are anemic, 10 percent of females in the reproductive years twelve to fifty are anemic, 1 percent of males in this age group are anemic, and 3 percent of people of both sexes over fifty are anemic. Also assume that 15 percent of the population is under twelve and 65 percent are aged twelve to fifty, with equal numbers in both sexes in all age groups. Find the overall incidence of anemia.

5. The sensitivity of a medical test is the probability of it giving a positive result when the patient actually has the disease. The specificity of a medical test is the probability of it giving a negative result when the patient doesn't have the disease. The proportion of people who have the disease in a particular community is 3 percent. The sensitivity of a test for this disease is 96 percent and the specificity is 92 percent.

    a. Find the proportion of positive tests that will occur when the population is tested.

    b. What is the probability that a person who tests positive has the disease?

    c. Repeat the question if instead of being used as a screening test in a population with a risk of disease of 3 percent the test is instead used for patients who are thought to be 95-percent certain to be suffering from the disease.

6. A judge initially believes that there is a 50–50 chance that the man before him is guilty. Forensic evidence is then produced that shows that the guilty man has blood group O. It is known that 50 percent of people in this region have blood group O and so does the man on trial. What should the judge now believe about the chance of the man's guilt?

7. A plane has crashed in an unknown location, but it was initially thought that there was a 40-percent chance (probability 0.4) that it had crashed over the sea and a 60-percent chance it had crashed on land. If it crashed on land there is a probability of 0.8 that a search will find it. If it crashed over the sea there is a probability of 0.3 that a search will find it. A search of land and sea has already been conducted and has failed to find the plane. What is the probability that it crashed on land? On sea?

8. a. How many ways can the numbers 1, 2, 3, 4, and 5, be arranged?

    b. Five people come into a room in which there is an armchair and a stool. How many ways can people be seated?

    c. Five people come into a room in which there are two identical seats. How many ways can people be seated?

   d. Two identical jobs are advertised and two women and three men apply. If there is no sex discrimination or affirmative action, what is the probability that the two women will get the jobs?

9. A class of ten students has five males and five females. Five students are selected to give class presentations through the semester. It turns out that only females are selected. The teacher is questioned about this but denies any sex bias.

   a. If the teacher selected students entirely at random, how often would it be that only girls were selected for presentation?

   b. Do you believe the teacher selected the students at random?

   c. The class is divided at random into two tutorial groups of five students. What is the probability that each tutorial group will contain students of one gender?

10. Consider the following table:

		First Method of Categorization		
		Category I	Category II	Totals
Second Method	Category A	5	0	5
of	Category B	0	5	5
Categorization				
	Totals	5	5	10

   a. What is the *p* value?

   b. Would you believe that there is an association between the two methods of categorization if the following are true (give reasons):

     i. Category I was not using a light while riding a bicycle at night. Category II was using a light while riding a bicycle at night. Category A was suffering a bicycle accident. Category B was not suffering a bicycle accident.

     ii. Category I was being born in the first half of the month. Category II was being born in the second half of the month. Category A was suffering a bicycle accident. Category B was not suffering a bicycle accident.

     iii. Category I was being a sports enthusiast. Category II was being uninterested in sports. Category A was suffering a bicycle accident. Category B was not suffering a bicycle accident.

   c. Repeat parts a and b with the number 5 in the table replaced by the number 2.

CHAPTER 4

Discrete Random Variables and Some Statistical Tests Based on Them

Don't be put off by the chapter title. The material here is no more difficult than the previous work. Random variables involve just a minor extension of the ideas in the previous chapter about probabilities of events that might occur. Random variables arise when the events that have various probabilities of occurring involve numbers. For example, say we examine three students and count the number of mobile phones they carry (assume that there is at most one mobile phone per student). We can get any of the numbers 0, 1, 2, or 3 for our result. The actual number will depend partly on the underlying average number of students in the particular social situation at hand who carry mobile phones, but the number will also depend partly on random chance. The numbers 0, 1, 2, and 3 in this experiment therefore have various chances attached to them.

   In this situation we use the letter X to denote the set of numbers 0, 1, 2, and 3 with chances attached, and we call X a *random variable*. In other words, X is used to denote the number of mobile phones we may discover when we examine three students, and the various values that X can take have certain chances attached to them. Since when we actually perform the experiment we have to get just one of the numbers 0, 1, 2, or 3, the sum of all the probabilities attached to the numbers must be 1. In the example here, X is called a *discrete* random variable because the values that X can take are values that are separated or discrete from each other. Discrete random variables generally involve just the whole numbers: 0, 1, 2, 3, 4, and so on. In rather different examples, the values that can be taken by a random variable may shade into each other. For instance, in some situations it may make sense for X to sometimes take a value such as 13.100, to sometimes take a value of 13.101, and to sometimes take any of the values in between the two previous values. In such cases, X is called a *continuous* random variable.

THE BINOMIAL RANDOM VARIABLE

Let us return to the experiment of examining three students for mobile phones. Say that the long-run proportion of students carrying mobile phones is θ. The proportion that don't carry mobile phones is $\phi = 1 - \theta$. (θ and ϕ are the Greek letters theta and phi, respectively. Many books use the English letters p and q in place of θ and ϕ, but the Greek letters are used here to avoid confusion with the p in the expression "p value.") The chance that the first, second, and third student we examined didn't have mobile phones would be $\phi \times \phi \times \phi$ by the law of multiplication; that is, the chance of getting 0 mobile phones would be ϕ^3. The chance of *no mobile phone & no mobile phone & mobile phone* would be $\phi \times \phi \times \theta = \phi^2\theta$. Similarly, the chance of *no mobile phone & mobile phone & no mobile phone* would be $\phi \times \theta \times \phi = \phi^2\theta$ and the chance of *mobile phone & no mobile phone & no mobile phone* would be $\theta \times \phi \times \phi = \phi^2\theta$, so overall the chance of exactly one mobile phone would be $3\phi^2\theta$. The coefficient 3 in the expression $3\phi^2\theta$ can also be deduced by reasoning that we have three slots and must choose one of them to be occupied by *mobile phone* with the remainder being occupied by *no mobile phone*. This choice of one from three can be done in $^3C_1$ ($= 3$) ways (further explanation is given in a later example). Similarly, the chance of exactly two mobile phones is $3\phi\theta^2$ and the chance of exactly three mobile phones is θ^3. The discrete random variable X here is then the set of these possible numbers 0, 1, 2, and 3, together with their associated chances ϕ^3, $3\phi^2\theta$, $3\phi\theta^2$, and θ^3. The discrete random variable X we have just described is a particular type known as the *binomial* random variable based on $n = 3$ and probability θ. This mouthful is abbreviated as $X \sim \mathrm{Bi}(3,\theta)$. If we were dealing with ten students, the number of mobile phones we could get could be any of the whole numbers 0 to 10, inclusive. The random variable would then be $\mathrm{Bi}(10,\theta)$. In general, if there were n students we would have the random variable $\mathrm{Bi}(n,\theta)$.

Of course, the idea doesn't just apply to students and mobile phones. It applies whenever there are n chances and we count the number of chances that actually come off. Formally, we say that the idea applies whenever there are n "trials" and we count the number of "successes." Extending the reasoning used in the case of three students, we can see that if $X \sim \mathrm{Bi}(n,\theta)$, the probability associated with X taking the value k is $^nC_k\theta^k\phi^{n-k}$. We write this as $\mathrm{P}(X = k) = {}^nC_k\theta^k\phi^{n-k}$, and these probabilities are sometimes called binomial probabilities. Of course, once we do the examination of the students we get a definite value: The number of mobile phones is no longer one of a range of numbers associated with various probabilities. But when we are about to do the experiment, the number of mobile phones we may find is X, a random variable.

OPTIONAL ▬▬▬▬▬▬▬▬▬▬▬▬▬▬▬▬▬▬▬▬▬▬▬▬▬▬▬

For example, say $X \sim \mathrm{Bi}(10,\frac{1}{4})$. Such a random variable would arise if we were about to perform the experiment of examining ten students and we somehow knew that the overall chance that any student had a mobile phone was ¼. Exactly the same random variable would arise if we were about to survey ten people about a political question and we somehow knew that the chance of a person giving a favorable response was ¼. Let us work out the probability that X takes the value 4 [in shorthand, $P(X = 4)$]. X takes the value 4 in a number of different circumstances. Let us denote p for mobile phone and n for no mobile phone. X takes the value 4 if we get any of the following results:

Result	Probability
ppppnnnnnn	¼×¼×¼×¼×¾×¾×¾×¾×¾×¾ = $(¼)^4 \times (¾)^6$
pppnpnnnnn	¼×¼×¼×¾×¼×¾×¾×¾×¾×¾ = $(¼)^4 \times (¾)^6$
pppnnpnnnn	¼×¼×¼×¾×¾×¼×¾×¾×¾×¾ = $(¼)^4 \times (¾)^6$
pppnnnpnnn	¼×¼×¼×¾×¾×¾×¼×¾×¾×¾ = $(¼)^4 \times (¾)^6$
⋮	⋮
⋮	⋮
⋮	⋮
nnnnnnpppp	¾×¾×¾×¾×¾×¾×¼×¼×¼×¼ = $(¼)^4 \times (¾)^6$

The question that arises here is how many possible arrangements there are of the letters p and n with four p's and six n's. To answer the question, imagine that we have ten slots laid out in a row. Into four of the slots we have to put the letter p, with the remaining slots getting the letter n. How many ways can we choose the four slots? This is the same question in effect as asking how many ways a teacher can choose four children out of her class of ten. In both cases the answer is

$$^{10}C_4 = \frac{10!}{4! \times 6!} = 210,$$

as explained in Chapter 3. Therefore, X takes the value 4 on $^{10}C_4$ or 210 occasions, with each occasion occuring with probability $(¼)^4 \times (¾)^6$. Therefore, $P(X = 4) = {^{10}C_4} \times (¼)^4 \times (¾)^6 \approx 0.146$.

END OPTIONAL

Note that the probability calculations for the binomial random variable use the law of multiplication, which is only valid when events are independent. Therefore the binomial random variable applies only when finding a mobile phone on one student does not affect our chances of finding a mobile phone on the next student. This assumption would not be reasonable if all students with mobile phones tended to stick together and so were likely to be selected together. In the case of asking a political question, the assumption of independence would not be reasonable if we simply asked ten people in one room their opinion. If people are in one room, they presumably have something in common; for example, they may be friends who share similar social backgrounds, which would tend to color their opinions in the same way. Although any one person in the room may have a probability of ¼ of being in favor, once we know that one person in the room is in favor, the chances are greater than ¼ that their friends in the same room will also have a similar opinion. In the extreme case, where people only ever share a room with others who agree on this particular political question, $P(X = 0) = ¾$, $P(X = 10) = ¼$, and $P(X =$ any number other than 0 or 10$) = 0$. In contrast, if there is independence between the responses of the ten people, $P(X = 0) = (¾)^{10} = 0.056$, and, $P(X = 10) = (¼)^{10} = 0.0000009$. When it is not correct to assume independence, the random variable X is not a binomial random variable.

The Use of the Binomial Random Variable in Statistical Tests: The Sign Test

Where Does the Name of This Test Come From?

The sign test is the main statistical test based on the binomial random variable. The sign test is used when we want to know if there is convincing evidence of an improvement or deterioration after some intervention has been applied to each of a number of individuals. The sign test gets its name because the only information that it uses is the sign of the change after an intervention (improvement can be denoted as "+", deterioration as "−"). An example will be used to explain the philosophical basis of the test and the calculation.

Example of the Use of the Sign Test

Say eight people were asked to compare the quality of two nights of sleep that they had where one of the nights of sleep occurred prior to listening to a relaxation tape and the other night of sleep occurred after listening to the tape. Say six out of the eight say that their sleep was improved after listening to the tape, whereas the remaining two people say the opposite. Does this constitute reasonably convincing evidence that the relaxation tape helps?

Some Philosophy behind the Sign Test and Hypothesis Testing in General

If we follow the same principles established in the case of Fisher's exact test, the way we use statistics to decide the issue is to weigh two ideas: (1) "The relaxation tape doesn't generally help sleep, it just looks that way in this particular group of people purely because of chance" or (2) "The relaxation tape does help sleep." As in the case of Fisher's exact test, we make our decision after calculating the chance referred to in option 1. If the chance is very small, it suggests that option 2 is a more reasonable conclusion.

However, some more discussion of these ideas is required here, and to facilitate this discussion some jargon needs to be defined. The idea in option 1, that the intervention made no difference, is called the null hypothesis and is denoted H_0. The idea in option 2, that the intervention improves the chances of someone having a good quality sleep, is called the alternative hypothesis and is denoted H_a. In the case of Fisher's test in the last chapter, we defined a main question and a subsidiary question. A similar idea applies in the case of the sign test and statistical tests in general. Using the terminology just defined, the main question can be rephrased as, "Which is more reasonable to believe, H_0 or H_a?"

In the case of the fired-women example of Fisher's test, H_0 was that there is no association between being a woman and getting fired and H_a was that there is an association. In the case of the sign test, H_0 is that the relaxation tape has no effect on the chance of someone's sleep after the tape being any better than their sleep before the tape. If H_0 is true and people are forced to decide which sleep was better, it is a 50–50 chance that they will say either "better" or "worse"; that is, the probability of "better" is ½. H_a, the alternative hypothesis, is that the probability of "better" is greater than ½.

However, the main question, "Is it more reasonable to believe H_a than H_0?" is not answered directly. Instead, a subsidiary question is asked: "If H_0 were true, how often would pure chance alone lead to results like ours that suggest that H_a is true instead?" If the answer is "hardly ever" (where traditionally "hardly ever" is taken to mean ≤ 0.05 of the time), then frequentist statistics tells us that we should leap from this answer to the subsidiary question to a response to the main question; namely, that there is convincing evidence for H_a. Conversely, if the answer is "quite often" (where traditionally "quite often" is taken to mean > 0.05 of the time), then frequentist statistics tells us that we should leap from this answer to the subsidiary question and respond to the main question by asserting that there is no (convincing) evidence for H_a. As has been repeatedly emphasized, this leap, if it is taken without thought and automatically using the traditional benchmark *p* value of 0.05, neglects common sense and inappropriate conclusions may result.[1]

Contrary to the impression given previously, frequentist statistics does not regard it as appropriate to assume that H_0 is true, simply work out the probability of getting the observed result, and then if this probability is "hardly ever" conclude that there is convincing evidence for H_a. Why not? The following example provides the explanation. Say instead of testing eight people with the relaxation tape we tested 1,000. Say also that 500 people said they slept better after the relaxation tape and 500 said they slept worse. Clearly, there is nothing in this result to suggest that the relaxation tape helps. However, let us calculate the chance of getting exactly 500 out of 1,000 under the assumption that the tape does nothing for sleep (i.e., the chance of a better or a worse sleep is ½). Using the terminology defined in the section on the binomial random variable, we have 1,000 trials, with each trial giving a success with probability ½. This is the chance that the random variable $X = 500$ where $X \sim \text{Bi}(1,000,$

½). Using the same ideas as before, this is the probability $^{1,000}C_{500} \times (½)^{500} \times (½)^{500}$. This probability turns out to be about $0.025 < 0.05$ = "hardly ever" (the probability is small because lots of values close to 500 out of 1,000, for example, values from 480 to 520 will all occur with comparable frequencies and so the probability of values "near" 500 out of 1,000 is shared out over quite a few numbers). The result of 500 out of 1,000 then "hardly ever" occurs, but this result certainly should not be taken as convincing evidence that the chance of a better sleep is anything other than ½.

The previous example shows that the probability question that must be asked needs to be more subtle than simply asking if the observed result would "hardly ever" occur if H_0 were true. Instead, the probability question asked is one that at first sight seems rather complex and obscure. The question is, "Assuming that H_0 is correct, how often by pure chance alone would we get results that look at least as much in favor of the explanation H_a as the results that have actually been obtained?" If the answer is "hardly ever," we would tend to believe that H_a is a better explanation. In the case of the relaxation tape and the eight people, six of whom slept better and two of whom slept worse, we would ask, "If the tape made no difference at all, how often by pure chance alone would six out of eight or seven out of eight or eight out of eight sleep better after listening to the tape?" If this answer is "hardly ever," we should conclude that we have reasonably convincing evidence for H_a. The logic behind this question is clearer if we state it in reverse. "If it would 'nearly always' happen that if H_0 were true chance would lead to us getting less than six out of eight people to say they slept better, then if our observed result was six out of eight this would be reasonably convincing evidence that H_a was true." Here we take the probability of "nearly always" to be 1 minus the probability of "hardly ever"; so, using the standard tradition of frequentist statistics, "nearly always" would mean with probability ≥ 0.95 or at least 95 percent of the time. In other words, if, assuming "better" or "worse" is a 50–50 choice, and if our probability calculations show that more than 95 percent of the time less than six people out of eight would say "better," then six out of eight is reasonably convincing evidence.

Put another way, we take the attitude before the experiment that the number that we get may lie in two regions. One region consists) of the most typical value that we would expect if H_0 were true and values close by. Values in this region will not make us change our minds about the assumed truth of the null hypothesis. The other region is a region of more extreme values that will make us change our minds. We are then interested in the overall probability of being in one or other region, assuming that the null hypothesis is true, but we are not interested in the probability of individual values.

What is our answer, then, for the particular result here of six out of eight sleeping better? We calculate the chance of at least six people out of eight saying they slept better when for each person there is a chance of ½ that they will say either "better" or "worse." Here we have a binomial random variable situation with eight trials and probability of success = ½. The chance of at least six out of eight is then the chance that the random variable $X = 6$ or 7 or 8, where $X \sim \text{Bi}(8, \frac{1}{2})$. Using the same ideas as before, this is the probability $^8C_6 \times (\frac{1}{2})^6 \times (\frac{1}{2})^2 + \, ^8C_7 \times (\frac{1}{2})^7 \times (\frac{1}{2})^1 + \, ^8C_8 \times (\frac{1}{2})^8 \times (\frac{1}{2})^0 = (\frac{1}{2})^8 \times (^8C_6 + \, ^8C_7 + \, ^8C_8) = 37/256 \approx 0.14$, which is more than the "hardly ever" benchmark of traditional statistics. This value, 0.14, is our p value. Put another way, if H_0 were true we would "quite often" get a result of six or more out of eight by sheer chance, even though the tape was of no benefit (here "quite often" is taken to mean a chance of $p > 0.05$). Traditional statistics therefore says that there is no (convincing) evidence that the tape benefits natural sleep, or that the evidence in favor of the tape is not statistically significant (or not statistically significant at the 5% level). Making (or inferring) a decision on an issue like this on the basis of such probability calculations is called statistical hypothesis testing or statistical inference.

We emphasize that the numbers in our experiment here are small. With small numbers, tendencies can easily be explained away as the effects of chance affecting just a few outcomes. The issue will be discussed further in Chapters 6 and 10. Although experiments will in practice generally involve larger numbers, there is no requirement in statistics to use sufficient numbers so that all tendencies of interest will usually be detected. The use of small numbers in the experiments discussed here, as well as simplifying calculations, highlights the common error of concluding that the experiment "proves" there is no effect when in fact there may well be a worthwhile effect but the numbers in the experiment were not sufficiently large for it to be unambiguously detected.

A person who was inclined to skepticism and had no knowledge of the benefits or otherwise of relaxation tapes could reasonably accept the conclusion here based on the 0.05 traditional benchmark. There is some evidence in favor of the tape if six out of eight say that it helped. But since this evidence could easily be mimicked by random chance, the evidence is not sufficiently convincing to entice a skeptical person to spend money on purchasing the tape unless that person was desperate enough to be prepared to waste money on something that may well not work.

The Problem of Ties

For the logic behind the sign test to work, we assume here that we don't allow people to simply say that the tape made no difference. We don't accept a "tied" decision. We insist that no two nights of sleep are exactly the same and our subjects have got to make a decision about which night was better. If any of our subjects refuse to make a decision, we ignore that person's response.

OPTIONAL

It is a philosophical question without an absolute answer whether this approach to ties is completely reasonable. For example, it may seem reasonable if we surveyed 1,010 people about the relaxation tapes and ten people refused to say better or worse, to then ignore those ten. But what if we surveyed 1,010 people and 1,000 of them said that they couldn't decide whether the tapes helped or hindered sleep: Would it be fair to entirely ignore all these responses and base our conclusions solely on the remaining ten who could make a decision? There are other possible philosophical approaches to dealing with ties. One approach involves breaking each tie arbitrarily and doing the *p* value calculation, then repeating the process for all possible combination of ways all the ties can be broken. The average of all the *p* values obtained is the final *p* value. In the pds computer program written to accompany this book the equivalent of the latter approach is used to deal with ties in the case of statistical tests discussed later in this chapter. However, in the case of the sign test, the approach of ignoring ties is used.

Another Sign Test Example

It is common for populations to consist of equal numbers of males and females. Now say that the sex ratio in grey kangaroos had not been studied before and so we decided to check whether the sex ratio was 50–50. If we caught four kangaroos and found that only one was male, would we be able to conclude that the sex ratio in kangaroos tends to favor females? There are two options: (1) The null hypothesis that the proportion θ is 0.5 or (2) the alternative hypothesis that the proportion isn't 0.5. To help decide which option 1 or 2 is more likely, we pretend at least for the sake of argument that the null hypothesis is correct. The outcome one male and three female is therefore regarded as a reflection of chance alone. We ask, "What sort of chance? How often would we get only one male instead of the two that we would most typically expect?" As explained previously, it is actually more sensible to alter the question slightly: "Normally we would expect two out of four; is the result one out of four so far away from two out of four that results this way out or further would rarely occur?" Using the theory of the binomial random variable we can calculate that we would get the numbers 0, 1, 3, or 4, $5/8$ of the time. So getting a number at least as far out as one out of four, from the "ideal" of two out of four, is not at all unusual. Therefore, if it was reasonable to start with the hypothesis $\theta = 0.5$, there still seems little reason to change our ideas.

If we had examined ten kangaroos and found only one male, we would go through the same reasoning. This time $P(X = 0) = (\frac{1}{2})^{10}$ and $P(X = 1) = {}^{10}C_1(\frac{1}{2})^1(\frac{1}{2})^9 = 10(\frac{1}{2})^{10}$. The calculations for $X = 10$ and $X = 9$ are identical to the calculations for $X = 0$ and $X = 1$. So assuming the hypothesis $\theta = 0.5$ is true, the chance of being as far as or further than our result of one out of ten from the ideal result of five out of ten is $2 \times (1 + 10) \times (\frac{1}{2})^{10} = 22 \times (\frac{1}{2})^{10} \approx 0.021$. This is the *p* value. We might then prefer to believe that option 2 is more believable than option 1. In other words, we might prefer to believe that the sex ratio in this kangaroo species is not 50–50, but is weighted in favor of females. Note again the reasoning: It might be 50–50, but if it is, a pretty small chance has come off to give us the data that make us doubt the 50–50 idea. Alternatively, there might be a good reason for the data. It might be this different from 50–50 not because a long chance came off, but because there is a good reason for it to be different, that kangaroos don't in fact have a 50–50 sex ratio. Which is the more plausible option is up to us, but we make a decision that is informed by a calculation. This is the calculation of the probability that, if the null hypothesis was true, by sheer chance we would obtain data at least as distant as our data from the most typical value of five out of ten, which the null hypothesis leads us to expect.

END OPTIONAL

Problems in Applying Bayes's Rule to the Sign Test

In the case of Fisher's exact test, after calculating our *p* value our next step was to apply Bayes's rule to incorporate our prior knowledge or common sense into the decision. We stated at the end of Chapter 3 that we would not formally apply Bayes's rule to further statistical tests because calculations would become too complicated. Let us see why. If the relaxation tapes are of some benefit, it would not be reasonable to expect that they would work every time. Chance factors may result in more disturbed sleep for some people after listening to the tape, even though the tape generally works most of the time. Application of Bayes's rule would require that we find out the probability of the result six out of eight under a range of scenarios from "the tape makes no difference" probability of ½ to "the tape is almost certain to work" probability of nearly 1. We would also need to judge how likely each of the scenarios in this spectrum was likely to be. Clearly, the Bayesian statistical approach in this context is going to be difficult. Instead, we "mentally" apply Bayes's rule. In other words, we apply a common-sense judgment to our result that the *p* value was 0.14 in the case of six out of eight sleeping better. As stated before, it would be reasonable for a skeptical person who has no knowledge in the area of the effects of relaxation tapes on sleep to conclude that the experiment has not produced convincing evidence. This would not be a reasonable conclusion if we were not so skeptical and perhaps common sense or other information suggested to us that the tapes could well be of some benefit. If the tapes were of no benefit the only way to explain a result of more than five out of eight is to say a 0.14 chance came off. If we already thought that the tapes might be of benefit, we would sensibly regard this as further evidence in their favor. It is important to note that the strength of our evidence is based directly on the *p* value. In other words, we judge the strength of our evidence by assuming H_0 is true and then asking how often pure chance alone would lead to outcomes that instead suggest H_a at least as much as our observed outcome does. The result "more than five out of eight" is not the immediate fact on which we base our judgment; it is the *p* value derived from this fact that is the direct evidence. The direct evidence here is that it would only happen 14 percent of the time that pure chance alone would result in an entirely useless insomnia treatment giving at least as much apparent benefit as was seen in this experiment.

Using p values rather than actual outcomes as a measure of the strength of evidence is not only logical, it also enables us to compare the strength of evidence when we are looking at outcomes that are otherwise hard to compare. For example, which provides the stronger evidence "six or more out of eight" or "three out of three"? Both results give roughly similar p values. Both outcomes therefore provide approximately equal strength evidence against H_0—something that would not be clear by looking at the outcomes themselves. Furthermore, without using p values to measure the strength of the evidence, we would have no way of knowing that four out of four can fairly be regarded as evidence that is twice as convincing as three out of three.

Note that the smaller the p value, the stronger the evidence against H_0. A very tiny p value generally means very strong evidence against H_0. It becomes less reasonable to attribute results to H_0 rather than H_a when the only way to attribute the result to H_0 is to argue that a very tiny chance came off that just happens by coincidence to make it look like H_a is true.

There is no correct conclusion. Statistics can help us make wiser decisions about situations where there is variability and uncertainty, but there is no method for always making a correct decision. Personally, if I had insomnia and some spare money, given the evidence of $p = 0.14$ I would buy the tape.

Using Numerical Results in the Sign Test

The sign test can be used in situations where the change as a result of an intervention is not classified simply as improvement or deterioration, but where actual numerical measures are available for performance before and after an intervention. For example, say the length of a night of sleep was somehow accurately measured for people before and after listening to a relaxation tape. Say that for eight people, A to H, the before and after results in hours and minutes were as follows:

Person	A	B	C	D	E	F	G	H
Before	5:37	6:24	4:22	6:53	3:19	5:07	6:48	7:09
After	5:18	6:47	7:01	6:46	7:31	8:08	7:51	8:11

We can summarize some of this information by stating that six people slept longer after the intervention and two people slept less. We can then perform the sign test exactly as before on this summary information. However, in doing so we are ignoring some of the original information. In particular, we are ignoring the fact that the people who had less sleep after the tape had only slightly less sleep, whereas the people who had more sleep had much more sleep. A more sophisticated test using all this information would provide stronger evidence against the null hypothesis.

On the other hand, if we are sure almost every reasonable person would be convinced by a p value of less than 0.001 and we get the result that ten out of ten people sleep longer after listening to the tape, there may be little need for more sophisticated tests using the actual numbers. If H_0 were true, a result of ten out of ten would occur only $(1/2)^{10} = 1/1024 < 0.001$ of the time (i.e., p value < 0.001). We have said that in this situation nearly all reasonable people would prefer to believe that the tapes work rather than believe that they don't work and that it just appears that way because a less than 1 in a 1,000 chance came off. Therefore, in this situation this simple sign test is all that is necessary to convince most people.

One-Tail and Two-Tail Tests

Consider a researcher who believes in the old saying "a healthy mind in a healthy body" and who believes that sport improves health. Such a researcher may wonder whether an after-school sports program improves academic performance. Let us put ourselves in the position of this researcher and say that we are prepared to believe in H_0, that the sports program has no effect on academic performance, and that we are prepared to stay with this belief unless we are convinced otherwise by a p value at least as small as the traditional benchmark value of 0.05. The alternative hypothesis H_a here is that sports programs increase academic performance. Now say that we conducted an experiment by funding eleven students to participate in an after-school sports program. Say that the result was that nine times out of eleven the student in the sports program improved in academic performance relative to the remainder of the students in the class. If H_0 were true, then we would have a result of less than nine out of eleven with probability 0.967 and a result of nine or more out of eleven would occur with probability or p value of about 0.033, since $^{11}C_9(1/2)^9 \times (1/2)^2 + {}^{11}C_{10}(1/2)^{10} \times (1/2)^1 + {}^{11}C_{11}(1/2)^{11} \times (1/2)^0 = (1/2)^{11}({}^{11}C_9 + {}^{11}C_{10} + {}^{11}C_{11}) = 67/2048 \approx 0.033$. Given our initial decision about the appropriate p value needed to convince us of H_a, we would therefore regard this as convincing evidence for H_a.

The initial viewpoint that a sports program might improve academic performance may seem unrealistic to some readers. Perhaps many people who dislike sports would believe that a sports program would, if anything, distract from academic pursuits and would therefore, if anything, hinder academic performance. Consider someone who believes that this hypothesis is possible, but who is yet to be convinced. If that person did not believe that a sports program could benefit academic performance but thought it was possible for a sports program to cause harm, then the sports program and academic performance experiment could only involve two possible hypotheses: either H_0 as before, that the sports program has no effect on academic performance, or H_a that a sports program decreases academic performance. The result in the experiment of nine out of eleven in favor of better academic performance for those in the sports program clearly is not ammunition in favor of an argument that sports programs detract from academic performance. What would a person conclude, who started off with the belief that most likely there was no connection between sports programs and academic performance (H_0), but if there was a connection it would have to be that there was a decrease in academic performance in those in sports programs (H_a)? Such a person would conclude that there is no connection (i.e., H_0 is true) and that pure perverse chance had worked to suggest a conclusion opposite to H_a. Even though a conclusion opposite to H_a is suggested by the evidence, the person evaluating the evidence will believe that the explanation for this evidence must be perverse chance and not a real effect, because that person believes that a real positive effect is simply not possible.

A more open-minded person might believe that while H_0 was a reasonable starting point, if there was some effect of the sports program on academic performance this effect could be in either direction. For this person, H_a would be that a sports program affects academic performance, without specifying the direction of the effect. To see how this affects our statistical reasoning, recall the philosophical basis for making a decision between H_0 and H_a. We make the decision after answering the question, "Assuming that H_0 is correct, how often by pure chance alone would we get results that look at least as much in favor of the explanation H_a as the results that have actually been obtained?" If the answer is "hardly ever" then we would tend to believe that H_a is a better explanation. For the open-minded person, a result of two out of eleven would be pointing toward H_a as much as a result of nine out of eleven (recall that H_a is now that academic performance is associated with the sports program but the association could be in either direction). Previously we calculated that 0.967 of the time we would get less than nine out of eleven (i.e., 0.033 of the time we would get a result at least as large as nine out of eleven). By symmetry, we see that 0.033 of the time we would get a result at least as small as two out of eleven (the chance of getting two heads in eleven tosses of a fair coin must be the same as the chance of getting two tails and so nine heads). Therefore, 0.066 of the time we get a result outside the range of three to eight out of eleven. The probability 0.066 is larger than the traditional "hardly ever" benchmark of 0.05, so if we are using this benchmark we cannot now say that we have convincing evidence against H_0. Put another way, if H_0 were true, 0.934 of the time we would get results that are closer to the 50–50 mark of five or six out of eleven than the outcome (nine out of eleven) here. However, 0.934 doesn't quite amount to "nearly always," so it is still "reasonable" to stay with H_0 even though the experiment gave an outcome outside the range of between three and eight out of eleven. The word "reasonable" is in inverted commas in the previous sentence because there is no reason other than convention to take "nearly always" as at least 95 percent of the time; as always, we should temper adherence to convention with common sense.

When H_a is that the intervention makes a difference but the difference could be in either direction, we say that we have a two-tail test. When H_a is that the intervention makes a difference in just one direction, we say that we have a one-tail test. The example here shows that whether a one-tail or two-tail test is appropriate and even the direction of the one-tail test are matters of subjective judgment. However, often it is clear that a particular intervention can do no harm but may or may not be of benefit. In such a case a one-tail test is appropriate.

The issue of one-tail and two-tail tests is often given considerable attention in standard statistical texts, as it is the only area in which some subjective judgment is needed in traditional statistics. However, it is a relatively trivial issue compared to the general issue of using subjectivity and common sense in statistics. Whether the traditional benchmark p value of 0.05 should be associated with just one end or spread over both ends of the range of possible outcomes is trivial compared to the issue of whether common sense would suggest an appropriate benchmark p value of the order of a billionth or close to 1. If common sense is giving us prior information that H_0 is almost certainly true, an exceedingly tiny benchmark p value might be appropriate, whereas if H_0 is quite likely to be incorrect, a benchmark p value close to 1 is appropriate. Such issues of common sense may then have far more effect on which outcomes are going to convince us about H_a than the issue of one-tail or two-tail tests.

Sign test calculations are often done by computer and the computer is generally programmed to give a p value assuming that a two-tail test is appropriate. If instead a one-tail test is appropriate, then it is often appropriate to halve the computed p value. However, some care is needed. In our example the computer would have given the p value 0.066 for our result "nine out of eleven," meaning that this is how often we would have got any of the results "nine or ten or eleven out of eleven, or two or one or zero out of eleven" by chance alone if H_0 were true. Half this 0.066 is indeed the required p value if H_a is that "sports programs increase academic performance," for in this case our p value is the probability of chance alone giving us results that look at least as suggestive of H_a as our result does, and only the results "nine or ten or eleven out of eleven" are in this category. However, if our alternative to H_0 was that the influence of sports programs, if any, must be bad, then our result of nine out of eleven of those in sports programs doing better gives almost no encouragement for us to believe in H_a. We would be more encouraged to believe in H_a, that sport is bad for academic performance, if we had got the result zero out of eleven improved, or one out of eleven improved, or two out of eleven improved, and so on, or even if we had got the result eight out of eleven improved. Results that are at least as suggestive of H_a as our result of nine out of eleven, then, consist of "nine or eight or seven or six or . . . or two or one or zero out of eleven." Calculation shows that such results occur $2{,}036/2{,}048$ of the time. In other words, the p value is 0.9941. Often it is correct to halve the computed two-tail p value when a one-tail test is appropriate, but some thought and common sense are necessary. If the results point in the direction opposite to that anticipated by H_a, the correct p value will be a number bigger than 0.5, not half the two-tail p value given by the computer.

The Sign Test and Pairing

The sign test is mainly used in situations in which there are "before" and "after" measurements on the same individual. However, the sign test is also used if there is a method of pairing individuals and one individual receives the intervention and the other individual doesn't. The "better" or "worse" comparisons of the outcomes between each of the pairs is then used in the sign test. For example, one child in each of a number of pairs of twins may be given a possibly beneficial treatment and the treated child can be compared with the sibling.

The Sign Test, McNemar's Test, and the Advantages of Pairing

The sign test can be used when there are pairs and where just one of each pair has the intervention, but the effect of the intervention does not have to be the classification "better or worse." Instead of "better or worse," the measure on each individual may be in the form of yes or no regarding some other attribute. In this situation, the sign test is given the name McNemar's test. Traditionally in this situation, p value calculations are derived using a different philosophical approach, but the calculations are equivalent to those discussed later in the section on large number of comparisons. McNemar's test will be discussed again in Chapter 8.

To explain McNemar's test, consider the following fictitious example. One thousand families are identified in which there are two children in early grades of schooling. One child in each family is selected to attend a reading appreciation course. Twenty years later, all 2,000 children, now in their late twenties, are followed up and asked if they had attended university. McNemar's test compares the answer of each person with their sibling. Where they both had attended universities or both not attended universities, their answers are regarded as ties and are ignored. Interest is then focused on those siblings where one had attended university and the other didn't. Under H_0, that the reading appreciation course was ineffective at encouraging university attendance, it should be 50–50 whether it was the sibling who attended or the sibling who didn't attend the reading appreciation course who later attended university. If out of the 1,000 pairs of siblings there were ten such pairs of siblings and in only one case was it the sibling who had attended the reading appreciation course who did not attend university, then our p value for a one-tail test is $P(X = 0) + P(X = 1) = (\frac{1}{2})^{10} + {}^{10}C_1(\frac{1}{2})^1(\frac{1}{2})^9 = 11(\frac{1}{2})^{10} = {}^{11}/_{1024} \approx 0.01$ [here, $X \sim Bi(10, 0.5)$]. If we thought that it was appropriate to use the traditional benchmark p value of 0.05, we would now state that we have convincing evidence that the course is effective in promoting attendance at university.

A point to notice here is that most pairs of siblings have been excluded from our calculations. There are 990 pairs where both siblings had the same experience of university attendance. This would reflect the fact that family attitudes toward education would be likely to have a much more important influence on the lives of young people and their ambition to attend university than the influence of a single reading enrichment course. Therefore, pairs of siblings having the same family background tended to either both go to university or neither go to university. If we had ignored the pairing we could have analyzed the situation using Fisher's exact test: There are two groups—those who attended the reading enrichment course and those who didn't—and two outcomes—university attendance or not. However, the analysis would be "muddied" by the 990 pairs or 1,980 people where family factors, not the reading enrichment course, were responsible for whether they had attended university. The experiences of the twenty people that the sign test or McNemar's test focus on would easily be accounted for in Fisher's exact test as chance fluctuations in large numbers. The point is that wherever possible we should control variability as much as we can in our experiments. In this example, with McNemar's test, this is done by pairing so that family factors are similar for both members of a pair. Pairing should be used whenever feasible to reduce natural variability that may obscure the effects of the experimental intervention. Unfortunately, though, pairing is often not possible.

The Sign Test and Testing Medians

The sign test can also be used to test whether figures that have been obtained come from a population with a known median value. For example, say the median house price in one town was known to be $67,000, and all seven houses that have been sold in another town (where records of median house prices are not available) have been sold for more than $67,000. Here, let H_0 be the hypothesis that the median house prices in the two towns are the same, and H_a be the hypothesis that the town without records has a higher median house price. Under H_0 there is a 50–50 chance (i.e., probability ½) that any house sold will be either above or below the median. Since seven houses have been sold above the median, the p value here is $(½)^7 = 1/128$, and we might well regard this as convincing evidence in favor of H_a.

Using the Sign Test When Numbers of Comparisons Are Large

Clearly the calculations involved in the sign test can be very lengthy if numbers are large. Say 65 out of 100 comparisons went one way: Is this reasonably convincing evidence that we are not dealing with a 50–50 situation? We are dealing with a binomial random variable X, with $n = 100$ and $\theta = 0.5$. For a two-tail test we want to know the chance that X is anywhere in the range 0 to 35, inclusive, or in the range 65 to 100. That is, we want

$$\sum_{k=0}^{35} {}^{100}C_k(0.5)^k(0.5)^{100-k} + \sum_{k=65}^{100} {}^{100}C_k(0.5)^k(0.5)^{100-k} = 2\sum_{k=0}^{35} {}^{100}C_k(0.5)^{100}.$$

This would be a horrendous calculation by hand or calculator. There is, however, a shortcut approximate method. This method is based on the normal random variable, which will be discussed later. However, such calculations are usually done by computer (in the pds program written to accompany this book, simply click on "statistical tests," then "sign test," then fill in the number of favorable and unfavorable comparisons and click "OK").

OPTIONAL

For example, let's say that we caught a sample of 100 kangaroos and found that there were 35 males. Our null hypothesis is again that the sex ratio is 50–50, with the alternative hypothesis being that it is some other value. The relevant probability calculation is to find out, under the assumption that there is a 50–50 sex ratio, how often we would get results as far out or even further from the most typical value of 50 males. We are dealing with a binomial random variable X, with $n = 100$ and $\theta = 0.5$. We want to know the chance that X is anywhere in the range 0 to 35, inclusive, or in the range 65 to 100. As shown, this chance is

$$2\sum_{k=0}^{35} {}^{100}C_k(0.5)^{100} \approx 0.004$$

Here, the calculation was done by computer. This is the required probability, the chance that even though the sex ratio is 50–50, when we catch 100 kangaroos at random we will actually find that the number of males is so far from the 50–50 mark that they are actually somewhere in the range 0 to 35 or 65 to 100. We have found that this chance is about 0.004, or 1 chance in 250. If we really did obtain such data we would probably think it more reasonable to believe that we were not dealing with a population with a 50–50 sex ratio, rather than believe that our out-of-kilter data are the reflection of a 50–50 sex ratio together with a 1 in 250 chance. To put it another way, if the null hypothesis had been true, at least 249 out of 250 times we would have got a value closer to 50–50 for the number of males than the result 35 males out of 100. Therefore, it is reasonable for the result we got, 35, to make us doubt the null hypothesis.

END OPTIONAL

A word of caution: For these calculations to be valid, the independence assumption underlying the binomial random variable must be satisfied. For example, say female kangaroos tended to stay together in an area and we sampled 100 kangaroos in that area. A result of 65 females might not be at all unusual, even though over the whole country the sex ratio is 50–50. Independence would not be satisfied in this situation. If as we sample, we find that there is a higher than expected proportion of females, it makes it likely that we are looking in an area where females congregate. If we continue to look in the same area, it is more likely than 50–50 that our next kangaroo will also be female.

Other Statistical Tests Based on the Binomial Random Variable

Sometimes we have some idea of what the long-run proportion θ is likely to be, based on theory or experience. We can then use our knowledge of binomial random variables in a statistical test to help us decide if our ideas are correct in the current case. For example, for theoretical reasons (Mendelian genetics) some genetic attributes called recessive attributes will occur with probability ¼ in each of the offspring of a particular mating pair. Say we observed eight offspring and five had the attribute. Does this mean that the attribute is not being passed on as a recessive in this case (i.e., does this mean that $\theta \neq$ ¼)? We note here that since, according to the laws of Mendelian inheritance for recessive genes, each offspring gets the attribute with probability ¼ and there are eight offspring, then on average we would expect two out of the eight to have the attribute, not the five that were observed. As in Chapter 3, it is not possible to answer the main question, "Is this attribute inherited with $\theta =$ ¼?" directly unless we use Bayesian statistics together with a priori estimates of how often alternative forms of inheritance occur. Instead, we provide an answer to the secondary question, "If θ was ¼, how often would we get results at least as far out as five out of eight?" and we use the answer to this secondary question to guide us in making a decision about the main question. For simplicity, we assume that H_a is that $\theta >$ ¼. Our p value, the probability of getting five or more out of eight when the proportion is actually ¼, is
$^8C_5(¼)^5 \times (¾)^3 + {}^8C_6(¼)^6 \times (¾)^2 + {}^8C_7(¼)^7 \times (¾)^1 + {}^8C_8(¼)^8 \times (¾)^0 = {}^{1,789}/_{65,536} \approx$
0.027. If we thought that it was appropriate to use the traditional benchmark p value here of 0.05, we would decide that we had convincing evidence of H_a. In other words, since it "hardly ever" happens that we get a result at least as high as five out of eight when the true proportion is ¼, then this result is convincing evidence that the true proportion is more than ¼. As noted previously, this statement may make more sense expressed in an opposite way: If the true proportion is ¼, we "nearly always" get results of four or less out of eight; since this didn't happen, we have convincing evidence that the true proportion is not ¼.

However, such statistical tests based on the binomial distribution with $\theta \neq$ ½ are not commonly used. The sign test, where $\theta =$ ½, is used more commonly. Taking $\theta =$ ½ represents the null hypothesis that better/worse or yes/no or male/female are 50–50 propositions. The example with $\theta \neq$ ½ was given to show how statistical tests can be based on any random variable appropriate to the situation.

IMPROVEMENTS ON THE SIGN TEST: USING ORDINAL INFORMATION AND THE WILCOXON SIGNED RANK TEST

Motivation

Consider the example of the night of sleep for eight people. The only way the numerical information was used was to classify outcomes as "better" or "worse." Clearly it should be possible to use the actual numerical changes in a more refined test. One such test that is often used (and is perhaps not quite so often used appropriately) is called the *t* test. This will be covered in the next chapter with some additional theory. In this section we will cover the Wilcoxon signed rank test, which doesn't use the numerical changes themselves, but uses their rank or order.

Explanation by Example

The test is best explained by an example. Consider again the example concerning the length of sleep with and without a relaxation tape. The results of hours and minutes of sleep are reprinted here with three extra rows. The first additional row is the difference between "before" and "after." The next row records the relative size or rank of the differences in the previous row, with a 1 representing the smallest difference and an 8 representing the largest difference. Signs are ignored in working out these ranks. The last row records the signs that were attached to the differences:

Person	A	B	C	D	E	F	G	H
Before	5:37	6:24	4:22	6:53	3:19	5:07	6:48	7:09
After	5:18	6:47	7:01	6:46	7:31	8:08	7:51	8:11
Difference	−19	+23	+2:39	−7	+4:12	+3:01	+1:03	+1:02
Rank	2	3	6	1	8	7	5	4
Sign	−	+	+	−	+	+	+	+

As with the previous statistical tests, we have two alternatives: H_0, that the intervention doesn't prolong sleep, and H_a, that the intervention does prolong sleep. Our primary question is, "Which is more reasonable to believe, H_0 or H_a?" In other words, our primary question is, "Does the relaxation tape really work, or does it just look that way in this particular group because of chance alone?" As previously, we do not actually ask this question directly, but instead we find a precise answer to the secondary question: "If H_0 were in fact true, how often would pure chance alone lead to results that are at least as suggestive of H_a as the outcome observed here?" In other words, the secondary question is, "If we are going to put the favorable results from the tape down to pure coincidence, what sort of coincidence are we dealing with?" If the answer is that we must be dealing with the sort of coincidence that "hardly ever" occurs, we conclude that that it is more reasonable to believe H_a. In the sign test, we judged whether the results were suggestive of H_a by p value calculations after simply counting the number of times we got the outcome "better" rather than "worse." With rank information we can do better. We look at the sum of all the ranks that are in favor of "better" or the sum of all the ranks that are in favor of "worse" (the total of these two sums is simply the sum of all the numbers between 1 and 8 [= 36], so once we know one sum we know the other). It would suggest that the intervention generally resulted in improvements if there were very few "worse" outcomes, and where "worse" outcomes did occur they were only slightly worse but when "better" outcomes occurred they were much better. This means that the intervention is suggestive of general improvement if the sum of the negative ranks is close to 0 or the sum of the positive ranks is close to the maximum value of 36. Let us return to our secondary question: "If H_0 were in fact true, how often would pure chance alone lead to results that are at least as suggestive of H_a as the outcome observed here?" If H_0 were true and we measure eight differences, the differences will receive the rankings 1 to 8 and each of these differences will be equally likely to be positive or negative.

We could get the sum of the negative ranks to be 0 if by chance all ranks were attached to a positive sign: Since under H_0 there is a 50–50 chance whether any difference is positive or negative, we see that we get the sum 0, $(½)^8$ of the time. We could get the sum of the negative ranks to be 1 if by chance all but the rank 1 were attached to a positive sign and the rank 1 was attached to a negative sign: Since under H_0 it is a 50–50 chance whether any difference is positive or negative we see that we get the sum 1, $(½)^7 \times (½)^1 = (½)^8$ of the time. Similarly we get the sum of the negative ranks to be 2, $(½)^8$ of the time. However, there are two ways of getting the sum of the negative ranks to be 3: Either we attach a negative sign to the rank 3 alone [this will happen by chance $(½)^8$ of the time], or as in our data, we attach negative signs to both the ranks 1 and 2 [this will also happen by chance $(½)^8$ of the time].

Altogether, we see that assuming H_0, that pure chance alone is operating, the sum of the negative ranks is 3 or less, $5 \times (½)^8 = {}^5/_{256} \approx 0.02$ of the time. This is our p value. If we thought that it was appropriate in the circumstances to use the traditional benchmark p value of 0.05 or any other benchmark above 0.02, we would conclude that we have reasonably convincing evidence that H_a is true. Again, the logic of the argument here is perhaps best understood if expressed in the negative. If H_0 (i.e., chance alone) was all that was involved, we would "nearly always" get a value for the sum of the negative ranks bigger than 3. Since we did not get a value bigger than 3, the alternative explanation— that the outcome is not just due to pure chance, but that the relaxation tape works—is a more convincing explanation.

Practicalities of Using the Wilcoxon Signed Rank Test

In the example, this test can be seen to be based on a random variable that can take values from 0 to 36 with various probabilities. However, it is actually quite difficult to work out all these probabilities, particularly when the numbers of comparisons are large. Unlike the case of the binomial random variable, there is no relatively simple formula for the probabilities here. The pds computer program written to accompany this book does the exact calculations when numbers aren't large; otherwise, it uses an approximate method. Other computer programs may use the more easily computed approximation even when numbers are small, and the approximation is quite inaccurate.

The p value calculated by the program assumes a two-tail test. In our previous example the program would give a p value of 0.04 with the sum of ranks of 3, since under H_0 the sum of negative ranks or the sum of positive ranks would be 3 or less about 0.04 of the time. It is appropriate here to divide this figure by 2 to get the p vlaue for a one-tail test. A one-tail test applies to our example because we are comparing sleep without any intervention with sleep after an intervention that might increase sleep but certainly wouldn't tend to reduce sleep. If instead we compared two different interventions that might promote sleep—say two different styles of relaxation tapes—a two-tail test would be appropriate. The two-tail H_a would be the hypothesis that one tape is superior to the other but we're not sure which. As in the case of the sign test, some thought is required. If a one-tail test is appropriate but the results point in the opposite direction to that expected by the alternative hypothesis, halving the computed p value would be wrong; the p value in such a case would be greater than 0.5. The computer program deals with the problem of any ties in the differences, using the principles described in an earlier section of this chapter, on pages 69–70.

USING ORDINAL INFORMATION IN UNPAIRED SITUATIONS: THE MANN–WHITNEY TEST

Motivation

Instead of comparing pairs of measurements, we often have measurements on unrelated individuals in two separate groups. We may then want to know if there is reasonably convincing evidence that the outcomes in the two groups are so different that the differences cannot reasonably be blamed on pure chance alone. We can already deal with this situation using Fisher's exact test. However, to use Fisher's exact test the outcome measured has to be to whether an individual belongs to one of two categories. If the outcome is a number, we could apply Fisher's exact test by categorizing the numbers as above average or below average. Clearly, though, we would be ignoring a lot of the information generated by the experiment if the only information we used was whether the outcome was above or below average instead of using the actual figures. There are methods of using the actual numbers directly in a statistical test. These methods will be discussed later. Here we will deal with the Mann–Whitney test. This test, like the Wilcoxon signed rank test, uses rank information rather than the actual numbers. Again, explanation will be by example.

Explanation by Example

Say we were interested in the relationship between gender and mathematical ability. I would guess that most people in our society would think that a reasonable starting point would be to stay with the belief H_0 that girls and boys are equally good at math unless we come across convincing evidence to the contrary. The alternative H_a would reasonably be taken to be that girls may be more talented at math than boys or vice versa. In other words, a two-tail test would be considered reasonable (a male chauvinist, however, may consider a one-tail test with H_a being that boys are superior to girls to be the appropriate alternative hypothesis). Let us now assume that we have some results that point to one gender being superior in math. As previously, our main question is, "Which is now more reasonable to believe, H_0 or H_a?" Once again, statistics cannot give a direct answer to this question and instead answers a subsidiary question like, "If H_0 were true, what is the chance that by pure coincidence we would get our results that just happen to make it look like H_a is true?" As discussed in the case of the sign test, the subsidiary question is actually a little more complicated than this. It is, "If H_0 were true, what is the chance that by pure coincidence we would get results that just happen to make it look at least as much in favor of the explanation H_a as the results that have actually been obtained?"

Let us now assume that we perform an experiment to test our ideas on gender and math ability and we obtain the marks of six boys and four girls of similar ages on the same math exam (we assume that our sample of boys and girls is a representative sample as discussed near the beginning of Chapter 2). Let's say that we obtained the following results:

Boys   63   72   45   48   27   51
Girls   96   61   88   86

These marks can now be combined and arranged in ascending order, with the prefix "b" representing the mark for a boy and "g" for a girl:

b 27   b 45   b 48   b 51   g 61   b 63   b 72   g 86   g 88   g 96

Dropping the marks, we can simply write bbbbgbbggg.

If H_0 is true, any arrangement of the six b symbols and four g symbols can occur. In fact, all possible arrangements of the six b symbols and four g symbols are equally likely. From Chapter 3 we know there are $^{10}C_4 = 210$ ways of positioning four g symbols into ten slots with the remaining slots being occupied by b symbols. Most of these arrangements will have the b symbols and the g symbols pretty well interspersed, though a few arrangements will tend to have most of the b symbols at one end and most of the g symbols at the other. These latter arrangements, though, are in favor of H_a. In particular, the arrangement that we obtained suggests that the girls are better than the boys. We now want to ask the subsidiary question, "If H_0 were true, what is the chance that by pure coincidence we would get results that just happen to make it look at least as much in favor of the explanation H_a as the results that have actually been obtained?" Before we can ask this question, though, we have to be able to define precisely what we mean when we say that have most of the b symbols at one end and most of the g symbols at the other. What we do is come up with a rule that gives a number that is 0 if all the b symbols are at one end and all the g symbols are at the other (bbbbbbgggg), gives a number that is small where there is a strong tendency for the b symbols to be at one end and the g symbols to be at the other (as in our current results), and gives a much larger number when the symbols are well interspersed (e.g., gbbgbbgbbg or ggbbbbbbgg). Once we have such a rule, we can sensibly ask the subsidiary question in this form: "If H_0 were true, what is the chance that by pure coincidence we would get an arrangement that using the rule gives a number at least as small as the number obtained from our results?"

The rule that has been devised is to look along the arrangement bbbbgbbggg and for each b symbol that occurs to count the number of g symbols to the left of it. Here the rule will give $0 + 0 + 0 + 0 + 1 + 1 = 2$. We can do the same thing interchanging the roles of the g symbols and b symbols. This will give $4 + 6 + 6 + 6 = 22$. The rule goes on to tell us to take the smaller of these two numbers, and traditionally we call this number U. The smallest out of 2 and 22 is 2, and so the rule finally gives us $U = 2$ as a measure of the extent to which most of the b symbols are at one end and most of the g symbols are at the other. By contrast, the arrangement gbbgbbgbbg gives $U = 12$ (counting the number of g symbols before each b symbol gives $1 + 1 + 2 + 2 + 3 + 3 = 12$ and counting the number of b symbols before each g symbol gives $0 + 2 + 4 + 6 = 12$, and 12 is the smaller of the two numbers 12 and 12). We now ask how many of the 210 possible arrangements of four g symbols and six b symbols give a value of 2 or smaller. Let us list some of the 210 possible arrangements together with their U values: bbbbbbgggg (0), bbbbbgbggg (1), bbbbbggbgg (2), bbbbbgggbg (3), bbbbbggggb (4), bbbbgbbggg (2), bbbbggbbgg (4). Trying out all 210 arrangements we find that only bbbbbbgggg, bbbbbgbggg, bbbbbggbgg, bbbbgbbggg, ggggbbbbbb, gggbgbbbbb, ggbggbbbbb, and gggbbgbbbb have a U value of 2 or less. This is 8 arrangements out of 210 possible arrangements. Under H_0, each arrangement is equally likely, so if H_0 is true, coincidence alone will lead to a result at least as much in favor of one gender as our results here only 8 out of 210 times. Our p value is then $^8/_{210}$, or 0.038. Using the traditional benchmark of 0.05 to be the upper limit of "hardly ever," we could say that if H_0 were true, then, although a result of U = 2 is possible, we would "hardly ever" obtain such a result and so it is more reasonable to believe H_a.

In practice, a declaration that one gender was intellectually superior in some respects could have major social implications and for this reason I personally would look for more convincing evidence before declaring to the world that girls are superior in math. After all, the evidence that we have could instead be explained by an 8 in 210 chance coming off. In coming to a conclusion, as well as weighing up the relative likelihoods of H_0 and H_a, it is also reasonable to take into account the costs of incorrect conclusions.

Sometimes an equivalent test to the Mann–Whitney test is referred to as the Wilcoxon rank-sum test, but this name can lead to confusion with the Wilcoxon signed rank test.

Practicalities of Using the Mann–Whitney Test

In the example, the test can be seen to be based on a random variable that can take values from 0 to 12 with various probabilities. However, as in the case of the Wilcoxon signed rank test, it is actually quite difficult to work out all these probabilities, particularly when the amount of data are large. There is no relatively simple formula for the probabilities here as there is for the binomial random variable. The pds computer program written to accompany this book does the exact calculations when numbers aren't large; otherwise it uses an approximate method. As in the case of the Wilcoxon signed rank test, the p value calculated by the program assumes a two-tail test and, with the precautions discussed on pages 75–76, it may be appropriate to halve this p value where a one-tail test is appropriate. Again, ties in the orderings are dealt with using the principles described in the section on ties in the sign test on pages 69–70.

OTHER DISCRETE RANDOM VARIABLES AND ASSOCIATED STATISTICAL TESTS

So far we have covered three discrete random variables. In other words, we have covered three situations in which there is a particular pattern of chances spread out over a range of whole numbers. These random variables are the binomial random variable and the discrete random variables associated with the Wilcoxon signed rank test and with the Mann–Whitney test. In fact, there are an unlimited number of random variables, since there are an unlimited number of ways of dividing up probabilities between various numbers. However, just a few discrete random variables are particularly useful and are given names. We will look at just one other discrete random variable.

The Poisson Random Variable

The binomial random variable applies to situations in which there are n independent "trials" with probability θ of "success" at each trial, giving a possible number of successes of 0, 1, 2, . . . , n. We can't get more than n successes in n trials. Often we deal with situations in which there is no theoretical upper limit to the number of successes. For example, we may know that on average there are three beetles of some particular type per square kilometer of habitat. We may then reason that since there are 100 hectares in a square kilometer, the chance of finding a beetle in any hectare is about 0.03. The actual number of beetles (rather than the average three) in the whole square kilometer would then be approximately given by a binomial random variable based on 100 lots of 0.03 chances (i.e., $X \sim$ Bi[100, 0.03]). Using the rules for the binomial random variable, we would be able to work out the probability that there were 0 or 1 or 2 or 3 or 4 or 5 or . . . or 100 beetles in the square kilometer.

But why stop at dividing the square kilometer into hectares? Working with square meters (1 million to a square kilometer), we would have the total number of beetles approximately given by $X \sim$ Bi(1,000,000, 0.000003) and could similarly work out the probability that there were 0 or 1 or 2 or 3 or 4 or 5 or . . . or 1,000,000 beetles in the square kilometer. Mathematically this process can be taken to the limit of an infinite number of infinitesimal chances but arranged so that the average remains three beetles per square kilometer. It turns out that the rule becomes

$$P(X = k) = \frac{3^k e^{-3}}{k!}$$

where e is the number close to 2.7183. Note that all the derivation of the rules that we have gone through up to now are understandable by anyone with some high school math and a logical mind. However, the reason for the rule just given requires further mathematics. This is true of most of the rules from now on, but the principle remains that these probability rules follow from basic logical principles.[2]

If, on average, we expect λ beetles per square kilometer the rule becomes

$$P(X = k) = \frac{\lambda^k e^{-\lambda}}{k!}$$

This random variable is known as the Poisson random variable (note that if k is 0 the rule will contain the expression 0!. This is taken to be 1. Also note that any number to the power 0, like λ^0, is also 1). The Poisson random variable applies to many situations in which the outcome may be any of the numbers 0, 1, 2, 3, 4, and so on without any upper limit. It applies at least approximately in a huge range of situations: number of gold nuggets in a patch of ground, number of typo errors in a book, number of hurricanes in a given period of time, number of customers walking into a shop, number of patients walking into a casualty ward, or number of telephone calls received in a given period of time. As it is derived from a binomial, it is implicitly assumed, as in the binomial, that the occurrence of one event is independent of the occurrence of all others. This is reasonable for short telephone calls. It would not be reasonable for long telephone calls, as one telephone call might interfere with the reception of other calls. It would also not be reasonable for beetles, unless they were solitary.

If we know that the probability of finding a certain number k of gold nuggets in a certain area is

$$\frac{\lambda^k e^{-\lambda}}{k!},$$

then the same principles that led to this rule show that the probability of finding k gold nuggets in twice this area is

$$\frac{(2\lambda)^k e^{-2\lambda}}{k!}$$

(provided the extra area is the same type of gold-nugget-bearing country as the original area). The principle extends in an obvious way, so that if we were dealing with, say, 2.56 times the original area, we would get a rule telling us our chances of getting any particular number of nuggets by replacing λ in the formula by $2.56 \times \lambda$.

As well as being used to deal with questions purely concerning probability, the Poisson random variable can be used in statistical hypothesis testing. For example, say long-term records show that a town experiences one destructive hurricane every fifteen years. How unusual would it be for the town to have at least three destructive hurricanes in the next ten years? As it stands, this is just a probability question and we will go through the workings shortly. However, in ten years time if there had been three destructive hurricanes we may be interested in whether this experience could fairly be regarded as convincing evidence for an increased hurricane frequency. Other knowledge tells us that any increased hurricane frequency would be due to the greenhouse effect. Our H_0, then, is that the greenhouse effect has not had a perceptible effect on hurricane frequency over the ten years, and H_a is that it has. Note that we have the same philosophical framework as before. What we really want is an answer to the primary question, "Are the increased hurricanes due to a change in climate (presumably) from the greenhouse effect, or have they occurred purely because of some unlucky coincidences?" We cannot answer this question directly. Instead, we answer the secondary question, "If we are going to put the frequent hurricanes down to unlucky coincidence, what sort of coincidence are we dealing with?" If the answer to this secondary question is that it is the sort of coincidence that "hardly ever" occurs, then we infer that the most reasonable answer to the primary question is that the frequent hurricanes are due to the greenhouse effect. Using the reasoning discussed earlier in this chapter, the relevant coincidence is not just the chance of three hurricanes, but the chance of three or more.

To start this problem, we need to know the λ to put into the rule for the chances for various numbers of hurricanes in ten years. Since we expect on average one hurricane per fifteen years, we expect on average 2/3 hurricane per ten years, so $\lambda = 2/3$. We want to know the chance of three or more hurricanes. We could use the Poisson probability rule to calculate the chance of exactly three, exactly four, exactly five, exactly six, and so on. To get the right answer for "three or more" we would have to keep going forever. However, "three or more" is the same as "not zero, one, or two." We can therefore find the probability of zero, one, and two, add them up, and subtract from 1 to give the probability of "three or more." The calculation then is

$$1 - \left[\frac{(2/3)^0 \, e^{-2/3}}{0!} + \frac{(2/3)^1 \, e^{-2/3}}{1!} + \frac{(2/3)^2 \, e^{-2/3}}{2!} \right] = 1 - e^{-2/3} \left[\frac{(2/3)^0}{0!} + \frac{(2/3)^1}{1!} + \frac{(2/3)^2}{2!} \right]$$

$$= 1 - e^{-2/3} \, (1 + 2/3 + 4/18).$$

A calculator shows the probability is about 0.03.

Note that the Poisson probability rule we have used may in fact not be appropriate, as like the binomial rule it implicitly assumes that hurricanes occur independently of each other. The Poisson probability rule would not be appropriate if, for example, as a result of an imaginary variant of the el Niño effect, we always had no hurricanes for forty-four years and then in the forty-fifth year we always had three. We would still have on average one hurricane every fifteen years, but we would get three hurricanes in a decade whenever the decade covered the forty-fifth year, which would happen by chance $10/45 = 22$ percent of the time (compare this with the probability of 0.03 or a 3% chance assuming hurricane numbers follow a Poisson random variable). In general, remember that calculations based on the Poisson random variable implicitly assume that the events, hurricanes, gold nuggets, whatever, occur at random, totally independent of each other. Often we can only make an educated guess about whether this assumption is reasonable. Presuming the assumption is reasonable, we go on to answer the primary question, "Do we believe that the excessive number of hurricanes is due to the greenhouse effect?" Here we also must make a judgment. Which is the more reasonable option to believe: (1) the greenhouse effect hasn't influenced the town's hurricane frequency, it just might seem that way as a result of a 3 percent chance coming off (H_0), or (2) the greenhouse effect has increased the town's hurricane frequency (H_a)? There is no right answer, only a judgment, but a judgment that is informed by the probability calculation that gave us the *p* value 3 percent.

Statistical tests based on the Poisson random variable are less commonly used than the sign test based on the binomial random variable, the Wilcoxon signed rank test, and the Mann–Whitney test, each based on their own special random variables. In fact, there is no name for the test given in the previous example other than "a statistical test based on the Poisson random variable." Nevertheless, the Poisson random variable comes up reasonably often in statistics.

SAMPLES, POPULATIONS, AND RANDOM VARIABLES

We have discussed random variables because of their role in statistical tests. They are also used in statistics in a somewhat different context. Random variables are used in the study of what a sample can tell us about a population.

In a sample, each value we obtain is chosen from a population. For example, we might have a sample of women recording the number of children each has, and with a sample of size 4 we might get a set of values $\{0, 1, 0, 3\}$. Each value is chosen from the values for the entire population of women. The collection of values for the entire population of women could be summarized as a collection of several billions of the numbers 0, 1, 2, 3, . . . , 30 (if 30 is the maximum number of children possible). Each of the numbers 0 to 30 occur in a certain proportion of the population. Choosing a value from this population is similar to choosing a value from a random variable which takes the values 0, 1, 2, 3, . . . , 30 with probabilities equal to the proportions in the population.

In the theory of statistics, we can think of the sample as a set of values obtained from a random variable rather than a population. This abstraction is useful because it allows some desirable simplification. A random variable is usually fully specified by a couple of numbers, or *parameters*, inserted into a short mathematical formula. The probabilities specified by an appropriate random variable might then give a close match to the proportions in the population. On the other hand, describing the population without this simplification would require us to estimate many proportions. In the example here we would need to estimate thirty-one proportions, the proportion of women who have zero or one or two or . . . or thirty children. Even a sample of thousands would contain very few women with more than twenty children, so many of the thirty-one proportions required to describe the population could not be accurately estimated. The method of using the sample to directly estimate the proportions would be inefficient because it fails to take into account that there is a pattern in the proportions to be estimated; for example, that past a certain number of children (the mode) the proportions of women with progressively more children tail off. The information in the sample is used more efficiently if we use the sample to estimate the couple of parameters in the mathematical formula that, at least approximately, describe all thirty-one proportions. Instead of dealing with the actual population, we are then dealing with a theoretical description of the population by a mathematical formula that specifies the chance that a woman chosen at random will have a certain number of children.

In the theory, then, we often think of the population as equivalent to a random variable. Being interested in the data because of what they tell us about the population is then the same as being interested in the data because of what they tell us about the underlying random variable that describes, at least approximately, that population. The theory underlying much of statistics is then based on what samples can tell us about the underlying random variables.

When we were dealing with summarizing data we defined the mean, or simple average, as one of the measures of where the data were centered, and the standard deviation as a measure of how spread out the data were. Now we are interested in the data for what they tell us about the population, and we have seen that this is equivalent to saying that we are interested in the data for what they tell us about the random variable that describes the population. It is therefore natural to ask how well the mean of the data and the standard deviation reflect where the underlying random variable is centered and how spread out it is. This raises the question of what we mean when we ask the question, "Where is a random variable centered and how spread out is it?" The definitions of center and spread for a sample can't be directly applied to a random variable. A random variable is a set of all the possible numbers that might occur, with various probabilities attached. The method of calculating the mean—adding all the values that actually have occured and dividing by the number of values—can't be applied directly to random variables.

Expected Value

The equivalent of the mean for a random variable is called the *expectation* or *expected value* of the random variable. It is the average of the numbers that might occur, but it is an average that also takes into account the probabilities of the various numbers occurring. The quick way of saying this is to say that it is a probability-weighted average of all the values that can occur. In mathematical symbols, if the possible values of the random variable are $x_1, x_2, x_3, x_4, \ldots, x_n$ and these values occur with probabilities $p_1, p_2, p_3, p_4, \ldots, p_n$, respectively, then the expected value is $x_1 \times p_1 + x_2 \times p_2 + x_3 \times p_3 + x_4 \times p_4 + \ldots + x_n \times p_n$. The expected value of a random variable is often given the symbol μ. For example, consider the random variable $X \sim Bi(3, \frac{1}{2})$. This takes the values 0, 1, 2, and 3 with probabilities $\frac{1}{8}$, $\frac{3}{8}$, $\frac{3}{8}$, and $\frac{1}{8}$, respectively. The expected value is then $0 \times \frac{1}{8} + 1 \times \frac{3}{8} + 2 \times \frac{3}{8} + 3 \times \frac{1}{8} = 1\frac{1}{2}$. For example, the expected number of heads when a fair coin is about to be tossed three times is $1\frac{1}{2}$. Likewise, the expected number of girls when a woman anticipates having a family of three children is $1\frac{1}{2}$.

Variance and Standard Deviation

The variance of a random variable is the probability-weighted average of the squared deviations from its expected value. If the values of the random variable that can occur are $x_1, x_2, x_3, x_4, \ldots, x_n$ and these values occur with probabilities $p_1, p_2, p_3, p_4, \ldots, p_n$, respectively, and the expected value is μ, then the variance is $(x_1 - \mu)^2 \times p_1 + (x_2 - \mu)^2 \times p_2 + (x_3 - \mu)^2 \times p_3 + (x_4 - \mu)^2 \times p_4 + \ldots + (x_n - \mu)^2 \times p_n$. For example, consider again the random variable $X \sim \text{Bi}(3, \frac{1}{2})$. The variance is $(0 - 1\frac{1}{2})^2 \times \frac{1}{8} + (1 - 1\frac{1}{2})^2 \times \frac{3}{8} + (2 - 1\frac{1}{2})^2 \times \frac{3}{8} + (3 - 1\frac{1}{2})^2 \times \frac{1}{8} = \frac{3}{4}$. The standard deviation of a random variable is the square root of the variance and is often given the symbol σ.

Expected Value and Standard Deviation of the Binomial and Poisson Random Variable

It can be shown that the expected value of a binomial random variable is $n \times \theta$. This is reasonable; it says that in some sense the average of what you would expect when you take n chances with each chance having probability θ of success is $n \times \theta$ successes. For example, if the sex ratio of some species was 50–50 ($\theta = 0.5$) and you examined ten individuals, you would expect five individuals to be female. If you looked at eleven individuals the expected number of females would be 5.5. The expected value doesn't have to be a whole number, even though the random variable may only take whole number values: The average number of children had by Western women is about 1.7, although no woman has exactly this number of children.

It can be shown that the expected value of a Poisson random variable is λ. This is reasonable, as λ was used as the long-run average in the derivation of the rule for a Poisson random variable. The details of the derivation are beyond the scope of this book.

It can be shown that the standard deviation of a binomial random variable based on n trials with chance θ of success each time and a chance $\phi = 1 - \theta$ of failure is $\sqrt{n\theta\phi}$. For a Poisson random variable it is $\sqrt{\lambda}$. Again, the details of the derivations are beyond the scope of this book.[3]

Best Estimates

We now return to the question of how well the mean and standard deviation of the data in a sample reflect the expected value and standard deviation of the underlying random variable. For example, is the sample mean the best estimate of the expected value that the sample can give us? Perhaps the median of the sample might be a better estimate? There is no answer to such questions that is true for all random variables. However, for most of the random variables mentioned in this book (the lognormal mentioned in Chapter 5 is an exception), the mean and standard deviation of the sample are the "best" possible indicators of the expected value and standard deviation of the underlying random variable.

OPTIONAL

This raises yet another question: What do we mean by "best"? There are a number of ways in which we can define "best." In many situations it turns out that the mean is the best estimate of the expected value of the underlying random variable according to all reasonable definitions of "best." The standard deviation

$$s = \sqrt{\frac{\sum_{i=1}^{n}(x_i - \bar{x})^2}{n-1}}$$

is the "best" estimate of the standard deviation σ of the underlying random variable in one sense, but the formula

$$\sqrt{\frac{\sum_{i=1}^{n}(x_i - \bar{x})^2}{n}}$$

gives an estimate that is "better" in another sense. In most situations the first formula is more appropriate. However, it is common for scientific calculators to have separate buttons for automatic calculation of both. The button for the first formula is commonly labelled s or σ_{n-1}, and this quantity is properly called the "sample standard deviation," and the button for the second formula is commonly labelled σ or σ_n. In practice, unless n is small there is very little difference between s and σ_n. The former is used to estimate standard deviation in the rest of this book.

Returning to the question of what we mean by "best," there are several criteria that are used in deciding which formula gives a better estimate. For example, one criterion is that if we were to take many samples and use the formula on each sample to make an estimate, then the average of the many estimates should be the true value. Another criterion is that if we make many such estimates, the scattering of these estimates around the true value should be as narrow as possible.

END OPTIONAL

SUMMARY

- Random variables are sets of numbers with chances attached.

- The binomial random variable applies when the numbers are the numbers of successes out of the number of independent trials.

- The most important application of the binomial random variable is to the sign test, where our starting point is that we have a 50–50 chance of success in each trial.

- The main application of the sign test is to situations in which we ask whether an individual is better or worse after an intervention.

- The sign test can also be applied to pairs where one of each pair gets the intervention. In this case it is sometimes known as McNemar's test.

- Ties are ignored in the sign test.

- A *p* value is not the chance of getting our particular result if the null hypothesis, H_0, is true. It is the chance that if H_0 is true we would get a result *at least* as suggestive of H_a as the result we actually obtained.

- A one-tail test applies if we believe the effect of the intervention could only possibly be in one direction. If the effect of the intervention could be in either direction, we have a two-tail test.

- The Wilcoxon signed rank test applies to the same situations as the sign test, but applies when we have more information than just "better" or "worse."

- The Mann–Whitney test applies when there are two groups and there is enough information to allow the outcomes of all the individuals to be ranked in order.

- Random variables are also convenient simplified representations of populations. The sample mean estimates the expected value of the random variable that represents the population. The sample standard deviation estimates the standard deviation of the random variable that represents the population.

In short, statistical tests covered so far are designed to help answer the question, "Is there a difference?" This question is asked in various contexts according to the source of the data and the different types of data:

Source of data	Dichotomous (e.g., better or worse) data	Numerical data
Two related measures (e.g., measures before and after an intervention on the same individual; measures on one twin who had an intervention and the other twin who didn't)	Sign test	Wilcoxon signed rank test
A single measure on two unrelated samples (e.g., measuring the same quantity on men and women)	Fisher's exact test	Mann–Whitney test

QUESTIONS

1. Ten asthmatics were asked to compare the effect of salbutamol and fenoterol sprays in relieving their asthma attacks. Assume that salbutamol and fenoterol are equal in terms of effectiveness and side effects in all people, so that the preferences expressed are simply the reflection of a 50–50 random choice.

   a. With this assumption, what is the probability that (i) zero people prefer fenoterol to salbutamol? (ii) one prefers fenoterol? (iii) two prefer fenoterol? (iv) three prefer fenoterol? (v) four prefer fenoterol? (vi) five prefer fenoterol?

   b. How low would the number prefering fenoterol to salbutamol have to be before you would believe that the results are not simply due to random chance but that salbutamol is a better drug? (Note there is no "right" answer.)

2. It has been claimed that a small machine that releases negative ions into the air in the bedroom enhances natural sleep.

   a. Ten people try the machine for one night and nine say that they slept better and one person says she slept worse. Do you believe that the presence of the machine enhances sleep?

   b. If thirteen people had tried the machine and nine had slept better, one worse, and three didn't find any difference, would you believe that the presence of the machine enhances sleep?

   c. If one-hundred ten people had tried the machine and sixty-five had slept better, thirty worse, and fifteen didn't find any difference, would you believe that the presence of the machine enhances sleep?

3. Say that in your local hospital at present there are twenty patients with broken legs. Sixteen are male and four are female. Do you believe that there is a general tendency for more men than women to be hospitalized with broken legs? (Assume that the sex ratio is 50–50 in your community.)

4. On two occasions, 1,200 people each run races against a stopwatch. On one of the occasions each person does some mental arithmetic exercises just prior to the race but does not do so on the other occasion. Say 700 people ran better after mental arithmetic. Assuming mental arithmetic is entirely irrelevant to physical performance, use a computer to calculate how often you would expect to get results at least this far out from the value of 600 out of 1,200 (the value most likely if mental arithmetic exercises are entirely irrelevant to physical performance). If you obtained this result, would you be reasonably convinced that prior mental arithmetic assists racing performance?

5. A tour company runs a twenty-four-seat bus. Knowing that 25 percent of people don't keep their booking, they only regard their twenty-four-seater as "fully booked" when they have twenty-eight booked on the tour. However, they will occasionally be caught out by this policy and have to pay compensation when more than twenty-four people on a "fully booked" tour do arrive to take their seats. In what proportion of "fully booked" tours will the company be required to pay compensation?

   a. Assume the tour company caters entirely to individuals who make their choice whether to proceed with the tour independently of each other.

b. Assume that all the tourists are traveling in family groups, with four in every family. Each family is independent of the other families, but if one member of a family can't come on the tour neither do any of the others in that family.

6. Students are randomly assigned to receive individual tuition from teacher A or from teacher B. The marks of students attending teacher A are compared with the marks of students attending teacher B on a standard test. The marks for the A group are 75, 84, 63, 42, 91, 87, 69, 73, 78, and 56. The marks for the B group are 89, 80, 47, 68, 96, 92, 78, 74, 61, 83, 79, and 88.

   a. Perform an appropriate statistical test to determine whether there is convincing evidence that the effectiveness of the two teachers is different.

   b. Would your conclusions change if instead of individual tuition the scenario involved two classes for the two groups of students taught by teachers A and B?

7. A group of people who regularly search the beach for lost valuables using metal detectors are asked to compare two different brands of metal detectors: brands A and B. The time in minutes for each person to find an item of value with each metal detector is recorded.

Person	1	2	3	4	5	6	7	8	9	10
Brand A	3	17	181	47	3	21	1	61	38	23
Brand B	19	15	176	93	6	48	26	108	57	29

Use a statistical test to decide whether there is reasonably convincing evidence that the two brands of metal detectors are not of equal effectiveness.

8. On average, a surgeon sees three cases of acute appendicitis each week. What is the probability that over the next week the surgeon will see zero, one, two, three, four, and seventeen cases?

9. On average a hematologist sees two new cases of acute myeloid leukemia each year. What is the average number of cases seen in three months? What is the chance the hematologist will see exactly four cases or four or more cases in the next three months? If, in fact, the hematologist does see four cases in the next three months, should he or she conclude that there is likely to be some new environmental factor responsible for these cases?

NOTES

1. This may happen if the amount of data examined by researchers isn't sufficiently large, the expected trend is not quite as strong as anticipated, or by sheer bad luck the data obtained turns out to be less convincing than it would normally be. The researcher may then have data that points in the appropriate direction but not in a completely convincing way. If statistics shows that chance alone could relatively easily explain the data ($p > 0.05$), then unthinking researchers would conclude that there is "no evidence" for the expected and observed trend. For example we may find statements that there is "no evidence" violent video games promote violent behavior, that there is "no evidence" that exercise promotes longevity, that there is "no evidence" that in-service training improves professional standards, and that there is "no evidence" that intensive fishing decreases fish stocks. Other examples in the same vein were given in on page 43.

2. Those interested in the derivation here should look up the derivation of the exponential function in an algebra textbook for senior high school or introductory university students (for example, Durell, *Advanced Algebra*, p. 128) and should also look up the Poisson distribution in an introductory probability theory book (such as Ross, *A First Course in Probability Theory*, p. 129). A bibliography can be found at the end of this book.

3. See Rice, *Mathematical Statistics*, pp. 111–113, or Ross, *First Course*, pp. 246–248, for the derivations.

CHAPTER 5

Continuous Random Variables and Some Statistical Tests Based on Them

Often data consist not of whole numbers but of values picked from somewhere over a continuous range. For example, data on the height of men would consist of values mostly picked from somewhere in the range 5′ to 6′6″. The important point is that the values are not generally separated from each other by whole numbers. If one man is 5′7.0000″ tall, the next talleer man could be 5′7.3062″ tall. Since values are not necessarily separated by any fixed amount, the possible range of values is said to be *continuous*.

When data were in the form of whole numbers they could be regarded as the outcome of a random variable that took whole number values. Such random variables—discrete random variables—were discussed in Chapter 4. In Chapter 4 we dealt with the binomial random variable, the Poisson random variable, and the random variables associated with the Wilcoxon sign rank test and the Mann–Whitney test. We used the theory of these random variables to work out probabilities concerning our data and this led to decisions about hypotheses.

Data that consist of values from anywhere on a continuous range can be regarded as the outcomes of a different sort of random variable: a continuous random variable. Continuous random variables are a little harder to understand. They still comprise a set of numbers with chances attached, but because there are an infinitude of close-by numbers, no single number can have a finite probability attached to it. Instead, we think of the density of probability around values of interest. Probability is smeared out over a range of numbers, but the density of the smear varies. For example, it makes no sense to ask the probability that a person is 170 cm tall, for nobody is exactly 170 cm tall if we take this to mean 170.0000000 cm tall and not 170.01386294 cm or any other number very close to 170.0000000 cm. However it does make sense to say that the probability density for human height at 170 cm is 0.05 per cm, meaning that *about* 5 percent of people are between 170 cm and 171 cm. The *about* is italicized because the 0.05 probability density applies only right at 170 cm. By the time we get to 171 cm the value may be something a little different, so the 0.05 probability per cm figure won't usually apply across the whole centimeter from 170 cm to 171 cm. The situation is a bit like speed: At a given instant a bicycle may be traveling at 5 meters per second, but if its speed is changing it will not cover exactly 5 meters in the next second.

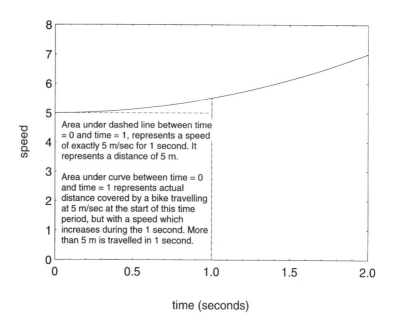

Graph of speed against time

The situation is perhaps clearer with a graph. Say we have a graph consisting of a horizontal axis along which the various possible heights of humans are marked, with a curve above describing the distribution of human heights (see page 99). This curve is called the probability density function for human heights if the area under the curve, between two heights of interest, is the proportion of people between those two heights. Equivalently, the area under the curve between two heights of interest is the probability that someone chosen at random from the population will be between the two heights.

We will deal with two related continuous random variables: the normal random variable and the lognormal random variable. The distributions of many continuous measurements that can be made in nature approximate either the distributions of the normal or lognormal random variables. The normal random variable arises when a large number of chance occurrences add together to give the whole. For example, height is determined by many factors added together. The factors include genes affecting the length of the various bones and nutrition before birth and during childhood, which in turn is the sum of many meals: All these influences add together to determine final height. The probability density function for human height is displayed here and shows a characteristic "bell-shaped" appearance. This is the shape of the probability density function of the normal random variable or the "normal probability density function" or the "normal curve."

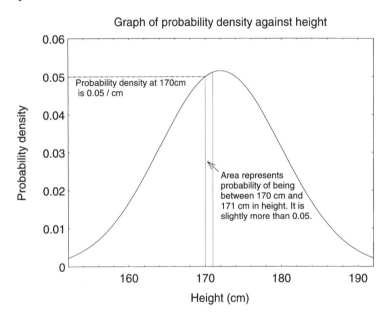

The lognormal random variable arises when a large number of influences are multiplied together. For example, the lognormal random variable describes the distribution of grains of sand on the beach. This random variable is the result of multiple splits from a larger rock. As discussed later, the final grain size is in some sense the result of multiplying together the effects of each split.

THE NORMAL RANDOM VARIABLE

Origin

It is a remarkable fact that the result of adding many, many, small chance contributions together gives almost the same underlying pattern of numbers and probability density as make up the normal random variable, almost regardless of the way chance is attached to each contribution. For example, consider an animal that is equally likely to get zero, one, two, or three meals on any given day. Let's say the animal will grow 0.0 mm that day if it receives zero meals, 0.1 mm if it receives one meal, 0.2 mm if it receives two meals, and 0.9 mm if it receives three meals. The end result in terms of the animal's length in adulthood after several years will be a particular pattern of lengths and probabilities. This pattern can be closely approximated by a bell-shaped probability density function of a normal random variable. However, any different pattern of meal probabilities and meal-related growth per day would give an almost identical pattern of lengths and probabilities for the adult animals. For example, if the probabilities of zero, one, two, or three meals in a day were instead $^1/_2$, $^1/_6$, $^1/_6$, and $^1/_6$, respectively, and the day's growth as a result of these meals was instead 0.0 mm, 0.3 mm, 0.6 mm, and 0.9 mm, respectively, the pattern of probabilities and lengths of the adult animal would be almost identical. The only differences would be in the position and spread of the approximating probability density function. No matter what the probabilities are for the various number of meals and no matter what the growth is for each meal, the end result would still be the bell-shaped probability density function of a normal random variable. This bell-shaped probability density function of a normal random variable satisfies a certain mathematical formula. The formula is a bit complicated and involves two parameters (a parameter is a dummy letter that can take the place of any particular figure in a formula, like the θ in the formula for the probabilities of the binomial random variable). It turns out that one of the parameters in the formula gives the location of the center, the expected value, of the normal random variable. This parameter is usually given the symbol μ, the traditional symbol for the expected value of a random variable. The other parameter gives the amount of spread of the probability density function as measured by the standard deviation. This parameter is usually given the symbol σ, the traditional symbol for the standard deviation of a random variable.

Notation

The shorthand for stating that the random variable X is a normal random variable centered on μ and with measure of spread (i.e., standard deviation) σ is $X \sim N(\mu, \sigma^2)$ (there is a slightly confusing convention to use the symbol for variance, σ^2, rather than the symbol for standard deviation, σ, in the notation here). Such a random variable X can take any value from minus infinity ($-\infty$) to infinity (∞), but values more than two or three times σ away from μ are very rare (95% of the time a value from a normal random variable will be within $1.96 \times \sigma$ of μ). If we are dealing with a situation in which the data that we have obtained consist of a scattering of values that are all particular outcomes of the same normal random variable $X \sim N(\mu, \sigma^2)$, we say that our data follow a normal distribution or that our data are normally distributed with mean μ and standard deviation σ. The expressions "*X* is a normal random variable," "*X* has a normal distribution," and "*X* is normally distributed" all have the same meaning.

The Central Limit Theorem

The remarkable fact that the end result of the addition of a huge amount of randomness generally results in a particular pattern of probability density entirely specified by just two parameters can be proved using advanced mathematics. This result is known as the *central limit theorem*. In fact mathematicians have been so fascinated with the central limit theorem that they have produced a number of different proofs. The simplest proofs apply when the chance effects are all of a similar tiny size and occur independently of each other. In the case of the length of an animal and the number of meals that it gets each day as it grows, the simple proof that the end result should be a normal distribution would apply only if two assumptions are true. These assumptions are that the chances concerning meals and growth each day are the same every day and that having a certain number of meals and growth one day has no effect on the chances for meals and growth the next day. There are more complex proofs that show that even when these assumptions are not strictly true, the end result is often still a normal distribution.

Central Limit Theorem Exceptions

The proofs are not valid in all cases. The exceptions include situations in which common sense would indicate that there would be limitations to the randomness. For example, in the case of the animals growing according to whether they had zero, one, two, or three daily meals, if there was extreme dependence between the meals for each animal so that any given animal always got the same number of meals every day, then clearly the adult animals would all end up as one of just four possible sizes (depending on whether we are dealing with an animal that got zero or one or two or three meals each day throughout its life) rather than ending up as a scattering of sizes following the pattern of the normal random variable. The proof, that the end result of addition of a lot of randomness is a normal distribution, is also not valid when some of the chance effects are much larger than others. For example, despite the contrived example in a previous section, human height is not distributed exactly as a normal random variable, because two chance effects—whether we are dealing with a male or a female, a child or an adult—have a much larger effect on height than how much a person got for breakfast on some particular day of their childhood. However, the height of adult humans of one particular gender is distributed much closer to the ideal of a normal random variable. If some of the chance effects multiply or combine in other ways rather than simply add, the end result may also not closely resemble a normal distribution.

Finally, as a mathematical curiosity, there are certain unusual random variables or patterns of numbers with probabilities to which the central limit theorem doesn't apply. No matter how many such random variables we choose independently and sum, we never approach the pattern of a normal random variable. One such random variable is known as the Cauchy random variable. It applies if we mount a laser pointer on a horizontal wheel near a long straight wall, spin the wheel at random, and record the position of the light on the (infinitely) long straight wall when the pointer comes to rest pointing to any part of the wall. While the pointer will more often point at parts of the wall close by, extreme values are not uncommon. It turns out that the extreme values are sufficiently common to prevent the mean of a number of such outcomes averaging out toward the more common values.

For most of the remainder of this chapter we will assume that in all the examples we are dealing with data that are normally distributed. The last section of this chapter discusses modifications required when we are dealing with the real world, where not all continuous data are normally distributed.

Calculations Based on the Normal Random Variable: The Standard Normal Random Variable Z

An unfortunate aspect of normal random variables is that there is no neat formula giving the exact amount of probability between any two points. However, approximate calculations of probabilities can be done using computers or statistical tables, and these calculations are simplified by using the following idea. The simplifying idea starts by noting that all normal curves have the same basic shape, they "look the same," and mathematically they are the same, except that they can differ in where they are centered, μ, and how spread out they are, σ. The amount of area (probability) under the curve between two points on one curve is exactly the same as the area between the two equivalent points on any other normal curve.

What do we mean by equivalent points? Say we wanted to know the probability of obtaining a value between 0 and 1 standard deviations to the right of the mean from a normal random variable centered on μ with standard deviation σ. In particular, say we knew that the mean height of women is 5' 6" and that the standard deviation is 3". Our question then becomes, "What is the probability that a woman chosen at random will have a height between 5' 6" and 5' 9"?" The answer is that the probability of such a value is the same as the probability of obtaining a value between 0 and 1 standard deviations to the right of the mean from any other normal random variable. In particular, if we use as a standard the normal random variable centered on 0 with standard deviation $\sigma = 1$, then the probability of being between 0 and 1 standard deviations to the right of the mean is the same as the probability of being between 0 and 1. The standard normal random variable is therefore defined as the normal centered on 0 with measure of spread (or standard deviation) 1. It is usually denoted Z, so in symbols, $Z \sim N(0, 1^2)$.

Probabilities concerning any normal random variable can therefore be related to probabilities concerning the standard normal random variable, and these probabilities can be obtained from tables or a computer. In particular, the pds program shows that the probability of a value between 0 and 1 from a standard normal distribution is 0.3413, so this is the probability that a woman chosen at random will have a height between 5' 6" and 5' 9". Equivalently, we can say that 34.13 percent of women are between 5' 6" and 5' 9". Note that the pds program doesn't give us this answer in a completely direct way. Instead, if we click on "statistical function," then click on "normal," and then type "1" in the box "z value," we get the answer that there is a probability of 0.8413 of obtaining a value less than 1 from a standard normal distribution. Equivalently, 84.13 percent of values are less than 1. The standard normal distribution is symmetrical around the value 0, so that 50 percent of the time we get a value less than 0. Therefore, if we deduct the 50 percent of values that are less than 0 from the 84.13 percent of the values that are less than 1, we find that 34.13 percent of values are between 0 and 1 (see Figure 5.1).

Sometimes it will be convenient to turn to tables of the standard normal random variable rather than turn on a computer program. These tables can be designed in various ways, but often show just probabilities of being less than a certain number of standard deviations to the right of the mean. Such a table is included as the appendix to this book. We may need to use symmetry to obtain the probabilities that we are interested in. For example, say we were interested in the chance that a woman chosen at random was between 5' 2¼" and 5' 5¼". Since the mean is 5' 6", we see that 5' 5¼" is ¾" below the mean. Since the standard deviation is 3", we see that this ¾" is ¼ or 0.25 standard deviations to the left of the mean. Likewise, 5' 2¼" is 1.25 standard deviations to the left of the mean. The required probability is then the probability of a value between 0.25 and 1.25 standard deviations to the left of the mean. This is the same as the probability of a value between −0.25 and −1.25 on a standard normal curve. By symmetry, this is the same as the probability of a value between +0.25 and +1.25 on a standard normal curve. In turn, this is the probability of being less than +1.25, take away the probability of being less than +0.25. To look up the probability of being less than +1.25, turn to the Appendix, go down the page until you come to the 1.2 label in the left-hand most column, then move across the 1.2 row to the column headed 5. The entry here is 8943, indicating that the required probability is 0.8943. Likewise, the probability of being less than 0.25 is 0.5987. The required probability of being between 0.25 and 1.25 is then the difference between these two probabilities,

Figure 5.1
Equivalent Points

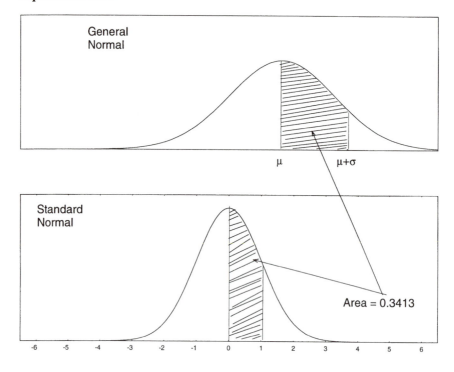

General
Normal

μ   μ+σ

Standard
Normal

Area = 0.3413

-6   -5   -4   -3   -2   -1   0   1   2   3   4   5   6

0.2956. In other words, if it is true that the mean height of women is 5' 6" and women's heights are normally distributed with a standard deviation of 3", then the probability that a woman selected at random is between 5' 2¼" and 5' 5¼" is 0.2956. Equivalently, 29.56 percent of women are in this height range (see Figure 5.2).

More generally, if we want to find the probability of a normal random variable $X \sim N(\mu, \sigma^2)$ taking a value between a and b, we find out how far each of a and b are from μ in terms of units of standard deviations, σ, and find the probability of being between each of these many units from zero in the case of the standard normal random variable. In particular, the distance between a and μ is $a - \mu$, and measured in terms of standard deviations rather than in the original units this distance represents

$$\frac{a - \mu}{\sigma}$$

standard deviations. Likewise, b is

$$\frac{b - \mu}{\sigma}$$

Figure 5.2

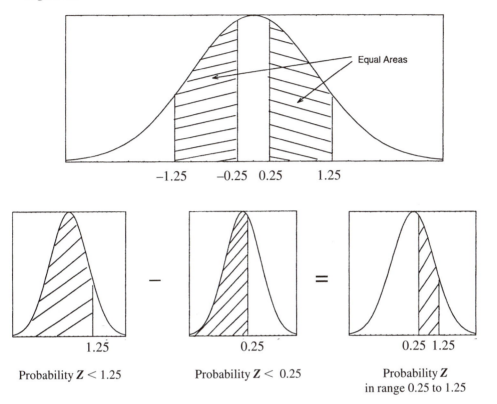

Probability **Z** < 1.25   Probability **Z** < 0.25   Probability **Z** in range 0.25 to 1.25

standard deviations from μ. The probability of *X* being between *a* and *b* is then the same as the probability of the standard normal random variable **Z** being between

$$\frac{a-\mu}{\sigma} \text{ and } \frac{b-\mu}{\sigma}$$

In effect, we deal with probability calculations by turning the normal random variable *X*, centered at μ and with standard deviation σ, into a standard normal random variable by taking away μ from all the values *X* and dividing by σ. In symbols,

$$Z = \frac{X-\mu}{\sigma}$$

where **Z** is the standard normal random variable.

Examples of the Use of a Normal Random Variable

1. What is the probability of obtaining a value of less than 1 from a standard normal random variable? It helps to refer to a diagram of the normal distribution, such as Figure 5.1. The probability required here is denoted by all the area under the curve to the left of +1 standard deviation. This value is given directly as 0.8413 by the pds computer program, as explained in the first example of the previous section. Similarly, looking up the value corresponding to z = 1.0 in the table in the Appendix shows the probability 0.8413. Therefore, there is about an 84 percent chance that a value obtained from a standard normal random variable will be less than 1.

2. What is the probability of obtaining a value of less than 1.645 from a standard normal random variable? The answer can be obtained directly from the computer, as previously described. If using the Appendix table, go across the row z = 1.6 until you get to the column headed 4. The entry here tells us that the probability of a number less than 1.64 is 0.9495. The next entry on the right under the column headed 5 tells us that the probability of a number less than 1.65 is 0.9505. We can then "interpolate" and conclude that the chance of getting a value that is less than the halfway point between 1.64 and 1.65 is halfway between 0.9495 and 0.9505, or 0.9500 (95%).

3. What is the probability of obtaining a value between −1.96 and +1.96 from a standard normal random variable? From the table we see that 97.5 percent of the time we get values below z = 1.96. Therefore, 2.5 percent of the time we get values above this. By symmetry, 2.5 percent of the time we get values below −1.96. Therefore, we get values in the range −1.96 to +1.96, 95 percent of the time.

4. If the distribution of the weights of adult humans is described by a normal random variable with mean 70 kg and standard deviation 10 kg, what proportion of people are between 50 and 90 kg? Asking for the proportion in this range is equivalent to asking the chance that any one person, chosen at random, will be in this range. If the weights that might occur when we are about to choose someone at random are denoted by X, the information in the question tells us that $X \sim N(70, 10^2)$. We convert X to Z by first taking away 70 and then dividing by 10, so that a weight of 50 is equivalent to a Z value of −2, and a weight of 90 is equivalent to a Z value of +2. In other words, the 50 kg is measured as 2 standard deviation units below the mean and the 90 kg is measured as 2 standard deviation units above the mean. Using the Appendix table as in example 3, we see that 1 − 0.9772 or 2.28 percent of the time we get a Z value above 2; likewise, 2.28 percent of the time we get a Z value below −2. Therefore, 95.44 percent of the time we get a Z value in the range −2 to +2. As explained already, this means that 95.44 percent of people have weights in the range 50 to 90 kg (this assumes, of course, that it is in fact true that weights are distributed as described here).

5. If the yearly minimum temperature in a town is normally distributed with mean 9°C and standard deviation 3°C, what is the chance that this winter the town will experience a temperature below freezing (0°C). The answer that follows uses the same logic as in previous examples, but we use this example to show how the logic can be expressed in more compact notation. The compact answer is that

$$X \sim N(9, 3^2), P(X < 0) = P\left(\frac{X-9}{3} < \frac{0-9}{3}\right) = P(Z < -3) = P(Z > +3)$$

$$= 1 - P(Z < +3) = 1 - 0.99865 = 0.00135.$$

In other words, there is a 0.135 percent chance or 1 in 740 chance. Put another way, if climate was not changing and yearly minimum temperature truly followed a normal distribution, we would expect on average one winter with subzero temperatures every 740 years.

THE LOGNORMAL RANDOM VARIABLE

As discussed before, the lognormal random variable is the result of the multiplication of a large number of chance effects. It can also be obtained by taking a normal random variable and using it as an exponent, $Y = e^X$. Conversely, the lognormal random variable can be changed into a normal random variable by taking logs.[1] This result is related to the fact that taking logs in effect turns multiplications (as in the multiplication of a large number of chance effects) into additions (as in the addition of a large number of chance effects). Like the normal random variable, the lognormal random variable is determined by two parameters, but its mean and standard deviation are more complicated functions of the two parameters. (Note the average of 2 and 4 is 3, but the average of 10^2 [= 100] and 10^4 [= 10,000] is not 10^3 [= 1,000]). This simple example shows that even though the average of X is μ and $Y = e^X$, we should not expect the average of Y to be e^μ.) The lognormal random variable can take values from 0 to ∞. A noteworthy feature is its very long right-hand tail. Earlier, one example of a lognormal random variable that was given was the size of grains of sand on a beach. In a sense, the size of a particle of sand depends on multiplying out the effects of all the splits in the original rock. For example, one particular grain of sand comes from a rock that was first split into equal halves. Then say that it was the right half that eventually gave rise to the grain, and that the right half was split one-third and two-thirds. Then say it was the two-thirds fragment of the right fragment that eventually gave the grain, and then the two-thirds fragment of the right fragment that eventually gave the grain was split into one-quarter and three-quarters. Then say it was the one-quarter that eventually gave rise to the grain, and so on many, many times. The size of the grain equals the size of the original rock × ½ × ⅔ × ¼ and so on many, many times.

A physical manifestation of the fact that the lognormal random variable has a very long right-hand tail is that there is an occasional sizeable pebble, even on a beach that consists mostly of fine sand. The lognormal distribution with this lack of symmetry—a long right-hand tail but a much smaller left-hand tail—is said to be positively skewed. In most of our work we will deal with normal random variables. Lognormal random variables are most easily dealt with by changing them to normal random variables by taking the log of the values; that is, the log of the size of the grains of sand on the beach are normally distributed.

HYPOTHESIS TESTS BASED ON
THE NORMAL DISTRIBUTION

As with our work on discrete random variables, the ultimate purpose is to make decisions: Is it reasonable to believe that relaxation tapes really increase the duration of sleep, or is it reasonable to believe that girls are better at math than boys? In Chapter 4 we made these decisions after making certain probability calculations. However, the calculations took into account only the order or ranking of the data values, not the actual values. In this chapter, instead of using just the ranking of the data values we will use the actual data values in making the relevant probability calculations and, hence, decisions. This approach, though, assumes that the values come from a normal distribution. This assumption may be justified by the central limit theorem, but sometimes the assumption will not be reasonable. Before coming to the examples of relaxation tapes and sleep or gender and math ability, we will start with a particularly simple situation. We will deal first with making a decision about whether just one particular data value comes from a known normal distribution.

Testing the Null Hypothesis That a Single Data Value
Comes from a Normal Distribution with
Known Mean and Standard Deviation: The z Test

It is unusual for a statistical test to be performed when there is just one data value. However, it is perfectly valid to do so. Little calculation is required and so it is easy to focus attention on the underlying philosophical principles. One example is given to illustrate the procedure and the underlying philosophy.

Say we were interested in physical anthropology and in particular we were interested in whether women from Madagascar were the same height as American women. I assume that I am dealing with an American readership that has no direct knowledge about the height of Madagascans, but who in the absence of any knowledge would assume that Madagascans have the same height distribution as Americans. Our null hypothesis, H_0, is just that: Madagascan women have the same height distribution as American women, so the average height of Madagascan women is the same as the average height of American women. Our alternative hypothesis, H_a, is that Madagascan women have a different height distribution than American women, so that the average height of Madagascan women may be either smaller or larger than the average American woman's height. Let us then assume that we know that the height of American women is normally distributed with $\mu = 5'\ 6''$ and $\sigma = 3''$. Let us also assume that we have met our first Madagascan woman, a tourist in the United States, and that she is 4′ 10½″. We need to also assume that this woman was chosen at random, without regard to height, from the Madagascan population. In other words, this Madagascan woman was as likely to be measured by us as any other Madagascan woman (if shorter people in Madagascar were for some reason more keen or more able than average-size Madagascans to obtain trips to the United States, the reasoning that follows would not be valid).

We start our calculation by assuming that H_0 is true, so we assume that the Madagascan woman has a height that has been chosen at random from a normal distribution with $\mu = 5′ 6″$ and $\sigma = 3″$. The woman we have just met is 7½″ or 2.5 (= 7½″/3″) standard deviations below the mean. Computer programs or statistical tables show that 98.76 percent of the time values chosen at random from a normal distribution will be closer to the mean than 2.5 standard deviations and only 1.24 percent of the time will they be at least as far from the mean as is our Madagascan woman (0.0062 is the probability of a value < -2.5 and 0.0062 is the probability of a value $> +2.5$, so $2 \times 0.0062 = 0.0124$ is the probability of being outside the range -2.5 to 2.5). We then have to weigh two options: "H_0 is true, Madagascan women have the same average height as American women, and our unusual result from our one Madagascan is explained by a 1.24 percent chance coming off," or "H_a is true, the height distribution of Madagascan women is different from that of American women, and this explains why our Madagascan woman does not have the height of usual American women." As in the previous chapter, we make our decision by comparing our *p* value of 0.0124 with some benchmark *p* value. As before, the benchmark *p* value describes the sort of coincidence that would be just sufficient to make us change our minds about whether we were dealing with the combination "H_0 with coincidence" or whether we were dealing with a real difference, H_a. If we believed that it is appropriate to use the traditional benchmark *p* value of 0.05, we would note that 0.0124 is less than 0.05 and conclude that Madagascan women are smaller than American women. Formally, we would state that we have statistically significant evidence (or the evidence is statistically significant at the 0.05 level) that Madagascan women are smaller than American women. It is entirely a matter of personal judgment whether we believe that the 0.05 benchmark *p* value is appropriate. However, it seems reasonable to me to use 0.05 here. It would be convenient not to have to store an extra fact in our brains that Madagascans have different heights to Americans, so we shouldn't discard H_0 because of evidence that could very easily be attributed to chance. On the other hand, it is clear that there is some racial variation in height in some instances so it would be silly to persist with the belief that there is no racial differences in height between Americans and Madagascans in the face of evidence that was difficult to explain by chance alone. With no great concern involved about the relative likelihoods of H_0 and H_a and no great concern about the costs of making an error in either direction, it would seem reasonable to me to use the traditional 0.05 benchmark p value and so I would conclude that Madagascan women are shorter than American women.

Many people learning statistics seem to be upset by this type of example: There is a feeling that a single case constitutes evidence that is far too flimsy to justify any conclusion. There are several points to be made here:

The first point is that it is not the number of cases that constitutes the strength of the evidence. The strength of the evidence is given by the probability calculation made on the assumption that H_0 is true. The result of this probability calculation is the p value, which tells us how easy or how hard it would be for coincidence alone to explain the result. A difference in heights with a p value of 0.0124 with one case is just as likely to be a real difference as a difference in heights with a p value of 0.0124 based on 100 cases. The value in taking 100 cases instead of one case is that smaller real differences will show up as being too big to be reasonably blamed on chance. If we had 100 cases we would get a p value of 0.0124 when the average height of the 100 Madagascans was only ¾" less than the average height of the Americans. The calculation here is explained later. Conversely, although this single case gave us reasonably convincing evidence that Madagascans differ from Americans, it would probably not have done so if Madagascans were only slightly smaller than Americans.

The second point is that even after taking on board the first point many people might still regard the evidence on which we are making a decision between H_0 and H_a as unreasonably flimsy. It could well be that we are concluding that Madagascan women are smaller than American women simply because the one Madagascan we happened to meet was one of the five out of every hundred Madagascans who have heights that are uncommon in both Madagascar and the United States. However, in this example we have used the same strength of evidence ($p = 0.05$ benchmark) that is used throughout the scientific world. In the real world we often have only flimsy evidence and we have to decide on that basis what is reasonable to believe. If we had more solid evidence we wouldn't need to use statistics. Unfortunately, many phenomena in the real world are subject to great variability and uncertainty, yet we still have to make decisions about factors that may have some influence on the phenomena. We have to do this even though the size of the influence is so small that it could easily be mistaken for the result of random variation. For example, many lifestyle and dietary changes and preventive medical interventions may only make a small amount of difference to the number of initially healthy people who are likely to die in five years. However, we will want to decide whether any differences in the death rate at five years are "for real" or just due to chance so we can decide what lifestyle and other interventions are worthwhile recommending. In this situation we are likely to be content to make a decision with evidence no more convincing (as assessed by p values) than our decision in the case of Madagascan women. To wait for the evidence to become much more convincing would require that we wait until most of the people trying out the different lifestyles that we are interested in have lived out their lives. It would require us to wait for a lifetime to get answers.

Some may continue to protest that statistical analysis is only valid if we have a representative sample, and they may argue that the height of one person cannot fairly represent heights in a whole nation. However, this protest is not valid. Statistical analysis requires representative samples, but the term "representative" here does not imply the usual full literal meaning of the word; it does not require that any individual in the sample be a typical representative of her nation. All that is required for our analysis to be valid is that each person in the sample is chosen entirely at random so that every person in the population has an equal chance of being chosen. If Madagascans are the same as Americans, then a randomly chosen Madagascan woman will have a 95 percent chance of being in the same height range as 95 percent of American women. If she is not in this height range, then using the benchmark p value of 0.05 we would regard this outcome under H_0 as too much of an unlikely coincidence and would rather believe that Madagascan women are different. We see that a sample of one is representative in the sense required by statistics, provided the single person has been chosen entirely at random.

It should also be noted that although we have only one piece of information about Madagascan women, we are comparing this information with complete knowledge of American women, knowledge that would have required a large number of measurements.

SAMPLING DISTRIBUTIONS

It is unusual for only one measurement to be available. More commonly a number of measurements are made of the same phenomenon. We measure the values from a sample of individuals obtained from the same population. We may measure the heights of a number of Madagascan women. We then want to use information from all these measurements to test the hypothesis that Madagascan women have the same height on average as American women. What sort of information should we use? The most obvious answer is that we should use the average of the measurements. Provided the underlying distribution is normal, mathematical theory agrees that this obvious answer is also the "best" answer.

We are then faced with a problem. We know that according to the null hypothesis individual measurements come from a particular normal distribution, and we have seen how to use this fact to obtain p values to help make sensible decisions about the truth of the null hypothesis. We are now dealing with an average, but it is the average of numbers that are subject to variation; hence, the average is subject to variation. We then need to know the distribution of the average of a number of measurements in order to use a group average to test a hypothesis in the same way as we used an individual measurement to test a hypothesis. Although the proof is beyond the scope of this book, the answer turns out to be simple (for proofs, see the probability and statistics texts in the bibliography). If the individual measurements are chosen independently of each other and come from a normal distribution with parameters μ and σ, then the average of n measurements comes from a normal distribution with parameters μ and

$$\frac{\sigma}{\sqrt{n}}.$$

Even if the individual measurements don't come from a normal distribution but come from some other distribution with expected value μ and standard deviation σ, then usually, if n is large (say 20 or more), the average of n measurements will usually have a distribution very similar to a normal distribution with parameter μ and standard deviation

$$\frac{\sigma}{\sqrt{n}}.$$

This result is a consequence of the central limit theorem.

This distribution of sample averages is known as a sampling distribution. In colloquial language, the sampling distribution tells us that the average of a whole lot of things that vary a fair bit about some central value will be something that varies about the same central value, but the amount of variation is reduced by a factor of the square root of the number of things used in calculating the average. This matches our intuitive idea that the average of a whole lot of measurements will be close to the true long-run average, but goes further and tells us how close to the true average we are likely to be. The fact that the factor by which standard deviation is reduced in going from individuals to groups is $\sqrt{n}$ requires some mathematics to prove, but it is related to the square root sign in the definition of standard deviation. Further explanation is beyond the scope of this book.

For some of the work that follows, translation of these statements into mathematical symbolism is useful, because once people get used to the symbolism, reasoning can be expressed more simply and precisely. In symbols, let X_1 describe the value we may obtain on our first measurement, and X_2 describe the value we may obtain on our second measurement, X_3 the third measurement, and so on. The result of averaging lots of random variables (sets of numbers with chances attached) is another random variable (set of numbers with chances attached). This average random variable is denoted

$$X = \frac{X_1 + X_2 + X_3 + \ldots + X_n}{n}.$$

If $X_i \sim N(\mu, \sigma^2)$ where the subscript i denotes the ith measurement and the X_i are independent of each other so that each of the n values are to be chosen independently from the same normal distribution, then

$$\overline{X} \sim N\left[\mu, \left(\frac{\sigma}{\sqrt{n}}\right)^2\right]$$

or, equivalently,

$$\overline{X} \sim N\left(\mu, \frac{\sigma^2}{n}\right) \quad \left[\text{Note that } \left(\frac{\sigma}{\sqrt{n}}\right)^2 = \frac{\sigma^2}{n}\right]$$

Therefore, noting the argument on p. 105, we have

$$\frac{\overline{X} - \mu}{\sigma/\sqrt{n}} = Z,$$

where Z is the standard normal random value. In other words,

$$\frac{\overline{X} - \mu}{\sigma/\sqrt{n}} \sim N(0, 1^2).$$

Briefly, we can say that if the individual measurements come from a normal distribution with parameters μ and σ, then the average of n measurements comes from a normal distribution with parameters μ and $\sigma/\sqrt{n}$. However, we need to keep in mind the important proviso in the more careful statement in the previous paragraph. The distribution with parameters μ and $\sigma/\sqrt{n}$ for the average of n measurements is only true if the individual measurements are chosen randomly and independently of each other. While this may seem an obscure requirement, failure to satisfy this requirement is in fact a major pitfall in practical statistics. If to save time we measured heights of a group of people in the same household instead of choosing each person in the sample for measurement independently, then it would not be so surprising to end up with a group of people who are all dwarves, since people in the same household tend to be related and may all have dwarfism for genetic reasons. If we chose people from separate households, obtaining a dwarf from every one would be most unlikely. The average height of household groups will be more variable than the average height of groups of individuals who have been selected independently of each other. It is not possible to predict the variability of groups where the individuals are not chosen independently.

Example of Use of Sampling Distributions in Calculations

Problem: Intelligence as measured by IQ testing is said to be $N(100, 15^2)$. According to the normal distribution table, a proportion of 0.1587 of all people, or 15.87 percent, have an IQ at least one standard deviation or 15 above the mean of 100. In other words, an IQ of 115 is the dividing line between the cleverest 15.87 percent of people and the rest of us. Now assume that for some reason many groups of four people have been formed in which each member of each group has been chosen randomly and independently of the others and the average IQ of the group of four is measured. What IQ level is the dividing line between the cleverest 15.87 percent of the groups of four and all the others? Equivalently, find the IQ level such that there is a 15.87 percent chance of the average IQ of the group being above this level.

Solution: If the standard deviation of individuals is 15, the standard deviation of the average of such groups of four is

$$\frac{15}{\sqrt{4}} = \frac{15}{2} = 7.5.$$

We are dealing with a normal distribution with mean 100 and standard deviation 7.5. The required value is one standard deviation above the mean. It is therefore $100 + 7.5$, or 107.5. If we repeat the question for the average IQ of a group of thirty-six individuals each randomly chosen independently of each other, the standard deviation for the group average is

$$\frac{15}{\sqrt{36}} = \frac{15}{6} = 2.5.$$

The answer is then that the required IQ level is 102.5.

If we repeat the question for a randomly chosen class of thirty-six children, no solution is possible. If the class has been chosen randomly, any child is as likely to be chosen as any other child in the population, so it is true to say that each child is chosen randomly. However, the children are not chosen independently of each other: They are all in the same class. If the first child we measured was well below average, it would be likely that some socioeconomic disadvantage may have contributed to his or her poor performance, and if so this socioeconomic disadvantage would be likely to be shared by other children in the same class who would probably come from the same locality. The average of classrooms of thirty-six children will therefore be more variable than the average of groups of thirty-six unconnected (i.e., independently randomly selected) children. Exactly how much more variable would be impossible to predict without further information.

Testing the Null Hypothesis That the Mean of the Data Comes from a Normal Distribution with Known Mean and Standard Deviation: The z Test Again

Most times we are interested in the possibility of changes that are not so convincing that they stand out with a single measurement. If we measured the heights of 100 Madagascan women and the average height was 5' 5¼", should we believe that Madagascan women are smaller than American women? Again we assume that we know that American women have heights described by a normal distribution with $\mu = 5'\ 6''$ and $\sigma = 3''$. Here the null hypothesis is the same as in the case of the height of a single Madagascan woman. In other words, we start with the belief that Madagascan women have the same height distribution as American women and that it is appropriate to maintain this belief and attribute any difference that we find to coincidence unless the coincidence is of the sort that "hardly ever" occurs. As in the previous example regarding the height of Madagascan women, I would regard it as reasonable to use the traditional benchmark *p* value of 0.05 or 5 percent as an appropriate specification of "hardly ever" in this situation. We will assume that each woman is selected randomly so that each woman chosen yields a value randomly selected from the normal distribution. We will also assume that the women have all been chosen independently of each other. From the previous section, we know that the average of 100 values from a normal distribution with $\mu = 5'6''$ and $\sigma = 3''$ will be a value from a normal distribution with $\mu = 5'6''$ and standard deviation $= 3''/\sqrt{100} = 3''/10 = 0.3''$. Our value of 5'5¼" is ¾" or 2.5 (= 0.75"/0.3") standard deviations below the mean. This is the same number of standard deviations as in the case of the single Madagascan woman of 4' 10½". As previously, computer programs or statistical tables show that 98.76 percent of the time values chosen at random from a normal distribution will be closer to the mean than 2.5 standard deviations. Since average heights would be at least as far out as 2.5 standard deviations less than 5 percent of the time, we conclude that there is evidence that Madagascan women are shorter than American women that is statistically significant at the 5 percent level.

We assess the strength of our evidence against the null hypothesis by finding out how small the chance is of obtaining results at least as extreme as ours by sheer coincidence. Therefore, we can say that the evidence we have obtained from 100 Madagascan women is exactly the same strength as the evidence we obtained from 1 Madagascan woman, since in both cases $p = 0.0124$. Why then bother measuring 100 Madagascan women when one would do? The answer is suggested by the examples. When measuring one Madagascan woman we would have a 50–50 chance of getting a woman as small as we did only if the average Madagascan woman was 4' 10½"; that is, 7½" smaller than American women. So with one Madagascan woman we would have a 50–50 chance of getting evidence against the null hypothesis that was less convinc-

ing than the evidence we obtained, even though the mean height of
Madagascans was 7½″ shorter than Americans. When measuring 100
Madagascans the same calculations apply when the height difference is only
¾″. In general, the more data values we have, the more likely we are to come
to correct conclusions about small but real differences.

One extra bit of jargon is useful here. In the calculations there are two different standard deviations that are relevant. There is the standard deviation of
individual values, σ. There is also the standard deviation of the average value
of a group of individual values,

$$\frac{\sigma}{\sqrt{n}}$$

To reduce confusion, the latter is given a separate name. It is called the standard error of the mean or simply the standard error.

Here is a second example, using some mathematical symbolism in place of
words. It is believed that radioactive particles in the atmosphere are a factor in
causing leukemia. With many poisons there is a safe dose, in the sense that a
small enough dose will cause no harm at all. However, radiation biology suggests that in the case of atmospheric radiation there is no safe dose: Small
increases in atmospheric radiation will cause a small increase in risk. However, this small risk applied to each of a large number of people will result in
small increases in leukemia incidence and hence deaths. The example then
concerns radiation levels after a minor nuclear accident. Say we knew that radiation levels over a city used to be normally distributed with a mean of 100 and
standard deviation of 30. Let's say that after an accident at a nearby nuclear plant
we took twenty-five measurements of atmospheric radiation and we obtained an
average level of 124. Does this provide evidence of increased radiation?

We follow the traditional statistical approach to answering this question by
starting with the null hypothesis H_0, that the postaccident radiation levels are
distributed in the same way as the preaccident radiation levels, with a normal
distribution centered on 100 and a standard deviation of 30. Note that the
reasoning we are using in this particular example would, for most fair-minded
people, not be credible. To convince those who would want to believe that the
accident has caused no pollution, we are saying that, in the first instance, we are
happy to pretend that there is no increase in atmospheric radiation as a result of the
accident even though we know that nuclear accidents often release radioactive
particles into the air. We are going even further and are saying that we will be
persuaded that there is a real increase only if whatever increase we actually encounter can't easily be put down to a coincidence. Perhaps there is some reason to
believe that in the case of this particular accident absolutely no radiation has
been released into the air or that a strong wind had blown all radioactive particles away from all population centers and all radioactive particles have been
deposited where they will not encounter humans until they have decayed. Oth-

erwise, our starting point, our null hypothesis or H_0, is an abuse of statistics and we really should not consent to such a null hypothesis. We note that the question, "Is there evidence that the nuclear accident has increased atmospheric radiation?" would be more rational if it were rephrased as, "Say we were to completely ignore the common sense that tells us that a nuclear accident is of itself evidence of increased atmospheric radiation and instead we were to look at the figures alone, would these figures give us convincing evidence of increased atmospheric radiation?" The null hypothesis that there is no increase in atmospheric radiation is employed here to illustrate the calculations, but its unreasonable use here also illustrates the widespread misuse of statistics.

In general, if the null hypothesis tells us that individual values X_i are $N(\mu, \sigma^2)$ (i.e., normally distributed with mean μ and standard deviation σ), then under H_0 the average of our data, $\overline{X}$, will be

$$N\left[\mu, \left(\frac{\sigma}{\sqrt{n}}\right)^2\right]$$

and we can judge the probabilities of various values of $\overline{X}$ by using the fact that

$$\frac{\overline{X} - \mu}{\sigma / \sqrt{n}}$$

will then have a standard normal distribution.

If we start with $X_i \sim N(100, 30^2)$, then the theory tells us that the random variable describing the average of twenty-five such measurements has standard error $30/\sqrt{25} = 6$, so $\overline{X} \sim N(100, 6^2)$. Then

$$P(\overline{X} > 124) = P\left(\frac{\overline{X} - 100}{6} > \frac{124 - 100}{6}\right) = P(Z > 4) = 1 - P(Z < 4) = 0.000032.$$

Another way of expressing this result is to say that an $\bar{x}$ value of 124 is equivalent to a z value of 4 and that z values at least as large as 4 occur only 0.0032 percent of the time. Here we have introduced new jargon. $\overline{X}$ as before is a random variable, as it is the set of values that we might obtain when we take our measurements and average them. Now we use $\bar{x}$ to represent the particular value of the random variable $\overline{X}$ that we actually do obtain when we take our sample and calculate the mean, and we use the lower-case letter z to represent a particular value obtained from a standard normal random variable (the latter is usually denoted Z). The value 0.000032 or 0.0032 percent that we get is again our p value. If we were to use $p = 0.05$ as our benchmark, we would reject H_0 that "radiation levels after the nuclear accident are unchanged." The increase in radiation is "statistically significant." Happily, on the figures provided, statistics tells us to reject a null hypothesis that common sense told us was inappropriate in the first place.

We note that a figure of 124 as a single measurement rather than an average would not make us reject H_0 at the $p = 0.05$ benchmark: It is only $^{24}/_{30}$ standard deviations above the mean, so it is less than the 1.645 standard deviations above the mean that is required for a one-tail test to be significant at the 5 percent level. Because we have twenty-five measurements, smaller variations from the long-run average of 100 show up as statistically significant. In colloquial language, the fact that taking averages allows us to reject the null hypothesis even though the changes in the values are not so marked reflects the fact that averages of twenty-five figures are five times less "wobbly" or variable or uncertain than a single figure. If the data average is some way away from what would be the central value according to the null hypothesis and the data average is not very "wobbly," the difference is likely to be real.

However, say the average value of the radiation level was $109 = \bar{x}$. This $\bar{x}$ value would be equivalent to a z value of

$$\frac{\bar{x} - \mu}{\sigma/\sqrt{n}} = \frac{109 - 100}{6} = 1.5,$$

and z values at least this big occur 0.0668 of the time. Since this is bigger than 0.05, the conclusion of the statistical analysis would classically be stated as, "There is no evidence that the nuclear accident has increased radiation." This conclusion would of course be inappropriate: We already know that there has been a nuclear accident likely to have emitted some radiation and we now see average measurements that have gone up so much above the previous average value of 100 that blaming the increase purely on variability means blaming the increase on a chance that occurs only 6.68 percent of the time. As previously, the conclusion here would be more sensible if it were stated as "looking at the figures alone, an average over twenty-five measures of 109 is not convincing evidence of an increase from the long-run average of 100 because the natural variability here could easily account for such an increase."[2] Of course, in this situation it is not sensible to look at the figures alone.

There is another important problem here. Our calculations assume that we have taken the average of twenty-five independent measurements. In environmental statistics it is often not completely true that each measurement is independent of all others. The most extreme example of lack of independence would arise if all twenty-five measures were performed on the one sample of air divided into twenty-five parts. While something like this might be necessary at some stage to check on the reliability of our measuring devices, it would only give us one effective measurement of the quality of the air. Less extreme examples would arise if many of the measurements were taken close together in time or space. Twenty-five such measures would give us less than the equivalent of twenty-five independent bits of information about the state of the air over all the time and space of interest. A lot more theory is required in order to know how to deal with such problems.

TESTING THE NULL HYPOTHESIS THAT THE DATA COME FROM A NORMAL DISTRIBUTION WITH KNOWN MEAN AND UNKNOWN STANDARD DEVIATION: THE t DISTRIBUTION AND THE SINGLE SAMPLE t TEST

Most times when we want to test an hypothesis about a mean from a normal distribution we do not have information about standard deviation. In our height example we may have good information that the true average height of American women is $5'6''$, but we may not know that $\sigma = 3''$. At this stage we recall that our null hypothesis is that the sample values that we have obtained from Madagascan women are distributed in the same way as the heights of American women. Therefore, if we assume that H_0 is true and we estimate the standard deviation of our sample, we have an estimate of the standard deviation of the heights of American women. We estimate this standard deviation σ by the sample standard deviation s in the data we collect (this can, of course, only be done if there is more than one measurement). We then replace the σ in our calculations by this s. In particular, we work out how far our average x is from the mean μ in terms of estimated standard deviations of the mean (i.e., estimated standard errors). How far $\bar{x}$ is from the mean μ is $\bar{x} - \mu$. The estimated standard error (or standard deviation of the mean) is $s/\sqrt{n}$, so the distance of x from the mean μ in terms of estimated standard errors of the mean is

$$\frac{\bar{x} - \mu}{s/\sqrt{n}}.$$

If H_0 were true and we had used σ instead of s, our result would be a z value (i.e., a value from a standard normal distribution). Because we use s in place of σ, the result is not a z value. It is called a t value. It is the result of a choice from a particular random variable known as the t random variable. The pattern of chances and values for t (i.e., the t distribution) will depend on the pattern of chances and values of the normal random variables that make up our sample. Since each of these normal random variables depend on parameters μ and σ, we might expect the t distribution to depend on μ and σ. However, t is defined so that it is standardized regardless of μ and σ, in the same way as a general normal random variable with parameters μ and σ can be standardized to the standard normal random variable Z. On the other hand, there is a different t distribution for samples of different sizes. If we have a sample that consists of just one value, we cannot have a t distribution, because from one value we cannot estimate the spread of values, s. If we have a sample consisting of just two values, we have one measure of spread. The value

$$\frac{\bar{x} - \mu}{s/\sqrt{n}}.$$

with $n = 2$ is then a value from a t_1 distribution. With a sample of three values there is in some sense two measures of spread and the value

$$\frac{\bar{x} - \mu}{s / \sqrt{n}}$$

with $n = 3$ is a value from a t_2 distribution. With a sample of n values there is in some sense $n - 1$ measures of spread and the value

$$\frac{\bar{x} - \mu}{s / \sqrt{n}}$$

is a value from a t_{n-1} distribution. The subscript $n - 1$ here is sometimes referred to as the "degrees of freedom," so instead of saying something has a t_{n-1} distribution we might say it has a t distribution with $n - 1$ degrees of freedom. When n is large, our estimate of variability s is based on a large number of measures of spread, and so s should be a fairly accurate estimate of σ. Therefore with n large there should be very little difference between

$$\frac{\bar{x} - \mu}{s / \sqrt{n}}$$

and

$$\frac{\bar{x} - \mu}{\sigma / \sqrt{n}}$$

Therefore, if n is large, the t_{n-1} distribution will be close to the standard normal distribution. Differences between the two distributions are very minor when n is more than about 20.

Let us examine again what we are doing when we calculate

$$\frac{\bar{x} - \mu}{s / \sqrt{n}}$$

We first calculate s, which is an estimate of how variable our data are. We then calculate $s / \sqrt{n}$, which is an estimate of how much variability we would expect in our sample mean given the variability of the individual data values. We then see how big the distance between the sample mean and the hypothesized true mean is $(x - \mu)$ in comparison to the amount of variability we can expect in our sample mean. This is the quantity

$$t = \frac{\bar{x} - \mu}{s / \sqrt{n}}.$$

Values outside a certain range will occur rarely. For instance, for a t_5 distribution, tables show that values outside the range −2.571 to +2.571 occur just 5 percent of the time. In other words, 95 percent of the time the sample mean is within 2.571 estimated standard errors of the true central value, μ. If we get a value outside this range we have to conclude that either we are dealing with a rare chance coming off or the hypothesized mean, μ, is incorrect. Traditionally, statistics tells us to use the 0.05 benchmark and so choose the second of these options, though, as has been repeatedly emphasized, we should always use common sense in deciding between such options rather than blindly following the traditions of statistics.

For example, say we measured the height of six randomly and independently selected Madagascan women and obtained the values 161, 169, 153, 165, 157, and 149 cm. Again, let H_0 be that Madagascan women have the same height distribution as American women, who let us say are known to have average height μ = 170 cm. As before, I believe it is reasonable in this situation to use the conventional benchmark *p* value of 0.05 in deciding whether to reject H_0. Calculations now give $\bar{x}$ = 159 and *s* = 7.483, so here

$$t = \frac{\bar{x} - \mu}{s/\sqrt{n}} = \frac{159 - 170}{7.483/\sqrt{6}} = -3.60.$$

Since 95 percent of the time when H_0 is true we would get a *t* value in the range −2.571 to +2.571 and we would "hardly ever" get a value of *t* at least as far out from 0 as −3.60, we reject the null hypothesis that Madagascan women have the same average height as American women. The sort of hypothesis test described in this example is sometimes called the single sample *t* test.

The Paired Sample *t* Test: Testing the Null Hypothesis That the Differences between Paired Data Points Comes from a Normal Distribution with Mean 0 and Unknown Standard Deviation

The statistical tests based on the normal distribution that have been described so far are not used that often. The main point in discussing them is to give some insight into principles underlying two common tests. The first of these two common tests is called the paired sample *t* test. This test is used in the same situation as the Wilcoxon signed rank test, which we discussed in Chapter 4. Both the Wilcoxon signed rank test and the paired sample *t* test apply where there is some pairing. Pairs of measurements are made. One measurement of each pair is made without the intervention and the other measurement is made with the intervention. Unlike the Wilcoxon signed rank test, the paired sample *t* test assumes that there is an underlying normal distribution. Both tests are used in obtaining an answer to the question, "Does some intervention affect the outcome?" The requirement of some pairing of the measurements applies if we have individuals and we measure outcomes before and after some intervention. Pairing also applies if we have pairs of similar individuals and measurement is made on the one of the pair who has received the intervention and the other who has not. The pairs of individuals could be pairs of twins, siblings, or even unrelated individuals who form pairs in that they are similar in age and other attributes. The difference between the measurement with and without the intervention is the value of interest. We recall that pairing is desirable to reduce the effect of variation from individual to individual, which may otherwise overwhelm the detection of real changes (see the discussion, pages 76 and 77, on the sign test, McNemar's test, and the advantages of pairing). The effect of individual variation is reduced because we are comparing each individual with himself or herself or with his or her pair. For example, the intervention might be supplying a relaxation tape and the value of interest would be the difference in the amount of sleep an individual got on the first night the tape was used in comparison to the amount of sleep the previous night. As another example, if we were interested in the effect of sports on academic performance, we might deal with human twins and the intervention might be an extra hour per day of sports for one of each pair of twins and the value of interest for each of the pair of twins would be the difference in their academic marks at school.

The necessary assumption of a normal distribution for the *t* test to be valid is often taken for granted. This assumption of a normal distribution is often a reasonable approximation, particularly where each measurement is a continuous value that is the end result of a large number of chance effects. Recall that the central limit theorem says that in many circumstances the end result of adding a large number of chance effects is a normal distribution. The average of a number of differences between two measurements will in turn be the end result of the addition of a number of chance effects. The issue of assuming a normal distribution is discussed further at the end of this chapter. However, in each situation, some thought is needed and if the assumption of a normal distribution is unreasonable, the Wilcoxon signed rank test should be used instead.

The null hypothesis H_0 in this paired situation will be that the intervention is entirely irrelevant to the outcome being measured: The differences in each of the pairs of measurements will be entirely due to chance alone. The differences will then be values randomly scattered about zero. In the paired sample *t* test, we are assuming that this random scatter follows a normal distribution. Our H_0 is therefore that the differences are values from a normal distribution centered on zero but with unknown standard deviation. In the previous section we dealt with testing whether data come from a normal distribution with known mean and unknown standard deviation. This is exactly the situation here. We test whether H_0 is reasonable by using the individual differences to find the amount of scatter in the system as estimated by *s*, the sample standard deviation. We then use this *s* to derive a measure of how much variability should be expected in the average of the differences. If there are *n* pairs of measurements we have *n* differences. The standard deviation of an average of *n* values is $s/\sqrt{n}$, the standard error, so this is the measure of the variability of the average of the differences. The null hypothesis is that the differences are randomly scattered about zero, but a sample of differences will not generally average out at exactly zero. Whether the average of our sample of differences is unreasonably far from zero is judged by comparing how far out from zero it is with how much variability could be expected of such an average. This is the quantity

$$t = \frac{\bar{x} - 0}{s/\sqrt{n}} \text{ or } \frac{\bar{x}}{s/\sqrt{n}} \text{ or } \frac{\bar{x}\sqrt{n}}{s}.$$

As before, if H_0 is correct, t is a value from a t_{n-1} distribution. If we get a value outside the range that is common for such t values, then we have to decide between two options. One option is that we believe that H_0 is still true and it is coincidence alone that has led to an average difference that is surprisingly far from zero given the amount of variability that appeared to be in the system. The other option is that H_0 is not true: The average difference in the pairs of measurements is not zero because the intervention does affect the outcome measured. As in the statistical tests discussed previously, we use the size of the coincidence given by the p value associated with the t value in making the decision between the two options. If we have a tiny p value, the only way to maintain our belief in H_0 is to argue that the results are the product of a very long coincidence. As always, we should also use common sense in deciding between the two options.

Readers may be a little confused about a small technicality here, in that we divided s by $\sqrt{n}$ but later used $n - 1$ in specifying the relevant t distribution. The n is used because the average of n measurements is less variable than the individual measurements by a factor of $\sqrt{n}$. The $n - 1$ is used in the calculation of s, the estimated standard deviation, because with the measurement of one difference we have no measurements of deviation of differences from the average difference, with measurement of two differences we have in effect one measure of deviation from the average difference, and with measurement of n differences we have in effect $n - 1$ measures of deviation from the average difference. These considerations lead, after some mathematics beyond the scope of this book, to the t_{n-1} distribution being relevant.

For example, let us return to the relaxation tape example, where we measured the amount of sleep had by eight individuals on a night without and on a night with the relaxation tape. The table of values is reprinted here for convenience. The differences are all given in minutes:

Person	A	B	C	D	E	F	G	H
Before	5:37	6:24	4:22	6:53	3:19	5:07	6:48	7:09
After	5:18	6:47	7:01	6:46	7:31	8:08	7:51	8:11
Difference	−19	+23	+159	−7	+252	+181	+63	+62

Here our average difference $\bar{x}$ is 89.25 and s is 97.52. Therefore,

$$t = \frac{\bar{x}}{s/\sqrt{n}} = \frac{89.25}{97.52/\sqrt{8}} = 2.589.$$

With eight values we have the equivalent of seven measures of deviation, and so if H_0 is correct this is a value from a t_7 distribution. A one-tail test is appropriate here, since if the relaxation tapes do anything for sleep they will lengthen it. Lengthening corresponds to a bigger value for $\bar{x}$ and hence a bigger t. A computer program will show that 98.2 percent of the time values from a t_7 distribution are in the range $-\infty$ to $+2.589$ and only 1.8 percent of the time are values 2.589 or more. Therefore, we are either dealing with a true H_0 and a coincidence of a sort that occurs less than 5 percent of the time, or H_0 is incorrect. If we use $p = 0.05$ as our benchmark p value, we reject H_0. It is appropriate here to note the discussion in Chapter 4 suggesting that it may be being unreasonably cynical to require such a small p value before believing in this H_a. However, the choice of a benchmark p value above 0.05 would not change our conclusion here. If our benchmark for rejecting H_0 is a coincidence that points toward the truth of H_a and would occur 5 percent or less of the time if H_0 were true, then the computer or tables show that this will correspond to a t value of $+1.895$ or larger.

The Independent Samples t Test: Testing the Null Hypothesis That Means of Data Values from Normal Distributions Are the Same

Often we deal with situations where there is no natural pairing and where before and after measurement is not possible. For example, we may be comparing the length of sleep in men and women. Although before and after measurement is desirable to reduce the effects of individual variability, we obviously couldn't very well measure the length of sleep in people of one gender, perform sex-change operations on them, and then measure the amount of sleep again.[3] The null hypothesis here is that on average there is no difference between the amount of sleep in people of either gender. We could deal with this situation by a Mann–Whitney test, but this test only uses the order of the data values and ignores the extra information that could be gleaned by using the actual values. To use the actual values we must assume that these values come from some probability distribution. The most common distribution in nature is the normal distribution. This is a consequence of the central limit theorem, so this is the distribution generally assumed (however, there is further discussion of this issue at the end of this chapter). The test, assuming that the data come from two independent samples and that there is an underlying normal distribution with unknown standard deviation, is referred to as an independent samples t test.

The calculations are a little more complicated than they are in the case of paired data, so instead of giving enough information to do the calculation using a calculator and tables, we will just give an outline of the calculation that the computer performs. Say that there are m values in one group and n values in the other. We will refer to the two groups as group 1 and group 2, respectively. Let us now examine the philosophy behind the calculations made by the computer.

There are four interrelated factors that complicate matters. First, the null hypothesis H_0 states that all $m + n$ values are drawn from the same unknown normal distribution. However there are three alternatives to the null hypothesis:

$H_a(i)$:　It may be that groups 1 and 2 have the same means but have different standard deviations.

$H_a(ii)$:　It may be that groups 1 and 2 have the same standard deviations but have different means.

$H_a(iii)$:　It may be that groups 1 and 2 have both different means and standard deviations.

Second, it is hard to imagine any influence that affects the mean that would not also affect the standard deviation, at least slightly. Conversely, almost any influence that affects the standard deviation would at least slightly affect the mean as well. Therefore, if the words "the same" in the definitions of $H_a(i)$ and $H_a(ii)$ are interpreted strictly, these options are implausible. The only plausible choice then is between H_0 and $H_a(iii)$. Nevertheless, it is conventional to ignore this consideration and regard all four hypotheses as feasible.

Third, most of the time our main interest is in whether the means are different regardless of any differences in standard deviations. We are therefore primarily interested in whether the data provide convincing evidence that means are different regardless of whether the standard deviations are equal. In other words, we start with either H_0 or $H_a(i)$, both of which state that the means are the same, and then see if there is convincing evidence against either of these hypotheses. In other words, we want to calculate a p value to see if there is convincing evidence that means are different, regardless of whether we provisionally assume standard deviations are equal or unequal, respectively. However, these two provisional assumptions lead to slightly different tests [note that if $H_a(i)$ is our provisional assumption, we are using $H_a(i)$ as a null hypothesis].

Fourth, the theory used to see if there is convincing evidence against $H_a(i)$ involves approximations and is less satisfactory than the theory used to see if there is convincing evidence against H_0. Classically, it is recommended that we use $H_a(i)$ (means equal, standard deviations unequal) as our null hypothesis if there is convincing evidence that H_0 is incorrect regarding equal standard deviations. This recommendation then requires a preliminary test known as an F test. This test looks to see if the standard deviations of the two groups are so different that chance alone would "hardly ever" lead to a difference in standard deviations at least as large as that observed.[4]

If calculations are being done by hand, we often perform just the statistical test to see if there is convincing evidence against H_0 in terms of different means.

The pds computer program gives the p value of this test as its one-line result, but like other computer programs performs all three statistical tests mentioned in the third and fourth alternatives. It gives the results of these tests on clicking on "moreinfo."

An additional approach suggests itself and is also performed by the pds computer program. If two groups are different in standard deviations they are not groups drawn from the same population and so are almost certainly at least slightly different in means. Therefore, both the test to see if there is convincing evidence against H_0 in terms of equal means and the test to see if there is convincing evidence against H_0 in terms of equal standard deviations are separate (in fact, independent) tests of whether we should believe the two groups come from the same population and so have the same mean. The minimum of the two separate p values from these tests is taken. An adjustment then has to be made. The adjustment takes account of the fact that by doing two tests we are giving ourselves two chances of rejecting H_0. If it were possible to do twenty tests to see if we should reject H_0 and our criterion for rejection was $p < 0.05$ on any test, then with twenty chances, each with probability 1 in 20 of occuring even when H_0 is true, we probably would end up incorrectly rejecting a true H_0 most of the time. In such a situation an adjustment is necessary so that the overall chance of rejecting a true H_0 will be 0.05. The same applies to a lesser degree when, instead of twenty tests, we perform two separate tests of H_0 (the test concerning equality of means and the test concerning equality of standard deviations). The pds computer program makes the appropriate adjustment to arrive at an overall p value. This approach can be summarized by saying that if the populations from which the samples are drawn are different in any way they are almost certainly at least a little bit different in all ways. This is an approach not favored by classical frequentist statistics. Instead, according to the classical approach, even if we must concede that there are differences in one aspect, such as standard deviations, we should wait to be convinced by the figures of differences in another aspect, such as means.

This discussion has concerned philosophical ideas underlying hypothesis testing in the case of two independent samples. We now deal with the issues in a more practical way. To test if there is convincing evidence against H_0 in terms of means, the computer averages the two sample standard deviations to give an overall best estimate of the standard deviation, usually denoted s_p where p stands for *pooled*.[5] Even if the samples are really both drawn from the same population, because of chance the means will not usually be exactly the same. However, theory tells us that in this case

$$\frac{\text{the difference between the two means}}{s_p\sqrt{\dfrac{1}{m}+\dfrac{1}{n}}}$$

should be a figure from a t_{n+m-2} distribution.

The components of this formula will be explained in a qualitative way, item by item:

The $n + m - 2$ reflects the fact that we do not know the true spread of values from this distribution, but we have in effect $m - 1$ estimates of spread about the mean from the first sample and $n - 1$ estimates from the second sample, giving $n + m - 2$ estimates of variation altogether. As before, the quantity $n + m - 2$ is often referred to as the degrees of freedom.

The $\sqrt{1/m}$ reflects the fact that the variation of an average of m values is $\sqrt{1/m}$th as variable as a single value.

The $1/m + 1/n$ reflects the fact that we are adding two sources of variation: variation in the average of the first m values and variation in the average of the remaining n values.

Overall, $s_p \sqrt{1/m + 1/n}$ is a measure of how much variability could be expected in the difference of the two means, so

$$\frac{\text{the difference between the two means}}{s_p \sqrt{\dfrac{1}{m} + \dfrac{1}{n}}}$$

is a measure of the size of the gap between the two means in comparison to how much variability should be expected in this gap.

Again there is a range of "likely" results for the

$$\frac{\text{the difference between the two means}}{s_p \sqrt{\dfrac{1}{m} + \dfrac{1}{n}}}$$

or t_{n+m-2} values. As before, we can decide on a range of values so that 95 percent or 99 percent or whatever percentage of the time we will get a value for t_{n+m-2} within this range. If we get a value outside this range we will say to ourselves we would "hardly ever" get such a value if the true means of the populations were really the same, and so we therefore no longer believe that the true means are the same (i.e., we reject the null hypothesis that the means are the same). As always, we should if possible use our own judgment as to what constitutes "hardly ever." Is this something that happens only 20 percent of the time or only 1 in 1,000,000 times? This is our benchmark p value. If we have no strong feelings on the issue, we may use the conventional benchmark p value of 0.05. The computer gives us the p value of results as extreme or more extreme than the value obtained and we compare this with our benchmark p value. The computer generally gives the p value assuming a two-tail test. This should be halved if a one-tail test is appropriate. This halving should not be done blindly. If a one-tail test is appropriate but results point in the direction opposite to those favorable to H$_a$, then the correct p value is 1 minus half the two-tail p value (see the discussion in the section on one-tail and two-tail tests on pages 75 and 76).

We now consider the possibility that we have decided that the standard deviations are different and want to know whether we should believe that the means are different as well. The underlying theory used here leads to similar calculations to those already discussed, but the value obtained for

$$\frac{\text{the difference between the two means}}{s_p\sqrt{\dfrac{1}{m}+\dfrac{1}{n}}}$$

turns out to be approximately a value from a $t_?$ distribution where "?" is worked out using a complicated formula. From this a p value is obtained and interpreted as before.

Overall, what we do can be summarized in simple language as follows. We make a judgment about whether the two group averages are so far apart that the difference between their averages cannot be reasonably put down to chance. We can judge this chance, the p value, from the overall amount of scatter in the values in the two groups.

For example, say that the hours and minutes of sleep were the same as given in the example used in the previous section on the paired sample t test. However, we interpret the values differently, so that the first row of numbers is the amount of sleep for a representative sample of eight women and the second row of numbers is the amount of sleep for eight men. In general, there would be no need for the number of women and the number of men to be the same. After selecting a file containing the data and clicking on "Independent samples t-test," the pds computer program performs the classical tests. The one-line output is that the p value is 0.024077. Clicking on "moreinfo" tells us that the p value for the test of equality of variances is 0.36639, the p value for the test of equality of means assuming equal variances is 0.024077, and the p value for the test of equality of means assuming unequal variances is 0.025756. The program then uses the logic described earlier based on the idea that the first two of these p values are both independent tests of whether the two groups come from the same population. This results in an overall p value of 0.047575. All p values assume two-tail tests. The program also gives further information about the average difference in sleep.

Which of these p values is pertinent depends on our precise philosophical approach to the situation. If the idea that we provisionally start with is that all $m + n$ values are drawn from the same unknown normal distribution and that we are prepared to do just one test of the equality of means of the two groups to see if we have convincing evidence against this idea, then the p value of 0.024077 is relevant. If the idea that we provisionally start with is that we are happy to concede that the standard deviations of the two groups are different but we would still look for convincing evidence before deciding that the means were different, then the *p* value of 0.025756 would be relevant. However, we should note that there is no compelling reason to concede that the standard deviations are different, for the computer also tells us that if two samples were drawn from two populations with the same underlying standard deviation, then 36.639 percent of the time the pair of sample standard deviations would be more different than the pair calculated here. If the idea that we provisionally start with is that all $m + n$ values are drawn from the same unknown normal distribution and that we are going to do two tests, one on means and the other on standard deviations, to see if we have convincing evidence against this idea, then the *p* value of 0.047575 is relevant.

It is left to us to come to our own conclusions. It might be reasonable to use the traditional 0.05 benchmark here, as H_0 is convenient, but the alternatives are not implausible. For me, it seems quite reasonable to start with the belief that men and women get the same amount of sleep on average, and the benchmark 0.05 may be appropriate because I would not want to be talked out of this belief by very weak evidence; nor would I require very strong evidence to talk me out of it. On the other hand, given the contentious nature of arguments between the sexes, it may be reasonable to want stronger evidence (in other words, a smaller *p* value) before making a pronouncement about differences between the sexes. If we accept the 0.05 benchmark it doesn't matter which of the *p* values we look at—0.024077, 0.025756, or 0.047575—they are all values that satisfy our criterion for rejecting H_0. If we accept the conventional benchmark *p* value of 0.05, then on the basis of this experiment we should now believe that there are differences in the length of time men and women sleep. However, is it indeed reasonable to now believe that there is a difference in sleep between the sexes? The figures we have analyzed could be explained by saying, "There is no difference, it just looks that way on these figures because a chance that would happen anyway a few percent of the time actually came off in our experiment." There is no objective answer to this question. The *p* value calculation informs our assessment of whether the difference in sleep between the two sexes is "for real" by telling us how hard it would be for coincidence to explain our results, but our decision is still subjective. Statistical tests cannot give absolute, objective results.

DATA TRANSFORMATIONS AND PARAMETRIC AND NONPARAMETRIC STATISTICS

The normal random variable results from adding many small chance contributions. It does not result from multiplying chance effects or from other interactions between chance effects. In nature, however, outcomes may be the result of a mixture of some random effects that add in combination with other effects that multiply or combine in more complicated ways, together with some random effects that are very large in comparison to most of the others. In such situations, the end result in terms of the pattern of randomness may not be particularly close to the pattern of a normal distribution. However, the normal distribution often seems to be a good approximation to many common situations where a continuous spectrum of numbers is possible, with values toward the middle of the range being much more likely than values toward either extreme. The common tests discussed in this chapter focus on the values likely to be taken by a sample average, and the central limit theorem tells us that such sample averages will generally give a much better approximation to a normal random variable than the original data.

Nevertheless, it is almost never possible to be sure that we are dealing with a situation in which the data values or even the sample mean come from an underlying distribution that is close to a normal distribution. If we were purists, we might avoid assumptions about the underlying distribution and use the statistical methods of the sort introduced in Chapter 4, which depend on just the order or ranks of the data and not on their actual values. These methods, such as the sign test, the Wilcoxon signed rank test, and the Mann–Whitney test, are valid regardless of the underlying distribution. However, since these methods ignore the actual values they do not use all the information in the data. This means that when there is an underlying normal distribution these methods are less efficient at detecting real differences. Since these methods do not depend on assuming that there is a particular distribution underlying the data and do not concern themselves with the normal distribution and its parameters μ and σ, these methods are often referred to as *distribution-free statistics* or, most commonly, as *nonparametric statistics*.

While a purist might want to always use nonparametric statistics at the cost of some efficiency, the opposite approach is to always use tests based on the normal distribution. When there really is an underlying normal distribution, the tests based on the normal distribution are more efficient because they take into account the actual data values, not just the ranks. However, these tests are only valid if there is an underlying normal distribution. For reasons already discussed, in many situations there is, at least approximately, an underlying normal distribution, and tests based on the normal distribution will give p values that would be close to the true p values. However, it is not valid to blindly assume that if we are dealing with a continuous random variable we are always dealing with a normal distribution. Such an approach is wrong. It can lead to very inaccurate p values and hence wrong decisions.

Checking Whether Data Come from a Normal Distribution

The widely used compromise approach is to check that it may be reasonable to assume that the data come from an underlying normal distribution before applying the statistical tests based on the normal distribution. This can be done in various ways.

Simple Display of the Data

If line plots, box plots, or histograms show that the data seem clustered more or less symmetrically about a central value with values toward the extremes being increasingly rare, the assumption of a normal distribution may be reasonable. Often, this is the only method of checking normality used in practice.

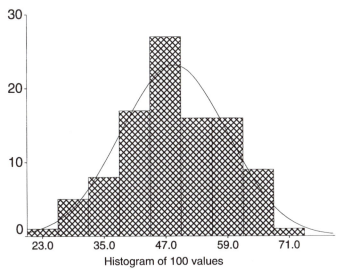

Histogram of 100 values

In the example here, the values were actually obtained from a normal distribution. Superimposed on the histogram is the curve of a normal distribution with the same mean and standard deviation as the data.

More Sophisticated Visual Indicators of Normality

There are more sophisticated visual indicators of normality that go under the headings of P–P plot and Q–Q plot and that provide a visual comparison of the spacing of data values compared to the spacing of values obtained evenly in probability terms through a normal distribution. Normality is indicated by an approximate straight line on these plots (see Figure 5.3).

OPTIONAL ▬▬▬▬▬▬▬▬▬▬▬▬▬▬▬▬▬▬▬▬▬▬▬▬▬▬▬▬▬▬▬▬▬▬▬

To explain Q–Q plots further, imagine we have chosen 100 values from a normal distribution. Place these values in ascending order. Going along this list should be like going along a normal distribution in the sense that halfway along the list of values should be roughly equivalent to halfway along the normal distribution in terms of probability. Similarly, the 86th value should be

Figure 5.3
Illustration of P–P and Q–Q Plots

Normal P-P Plot of Normal Random Variable

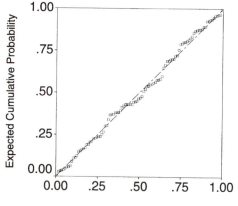

100 values chosen randomly from a normal distribution

Observed Cumulative Probability

Normal Q-Q Plot of Normal Random Variable

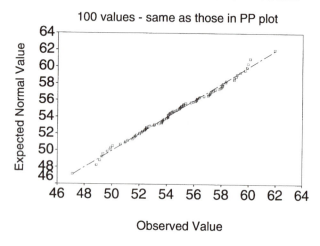

100 values - same as those in PP plot

Observed Value

roughly equivalent to 86 percent of the way along the normal distribution in terms of probability, or in general, *y* percent of the way through the values should be roughly equivalent to *y* percent of the way along the normal distribution in terms of probability. We can use this principle to find a value from the normal distribution, z_i that corresponds in terms of amount of probability behind it, to each data value x_i. If the data come from a normal distribution and we subtract an estimate of the mean from the data values and divide by the

Figure 5.3 (*continued*)

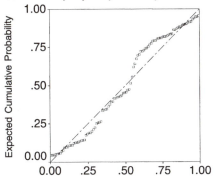

Normal P-P Plot of Uniform Random Variable

100 values equally likely to be anywhere in an interval

Observed Cumulative Probability

Normal Q-Q Plot of Uniform Random Variable

100 values - same as those in PP plot

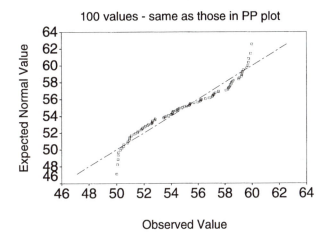

Observed Value

estimated standard deviation, we will have roughly converted the data values into values chosen from a standard normal distribution. Therefore, there should be a rough match between values of

$$\frac{x_i - \bar{x}}{s}$$

and corresponding values of z_i. For example, the 86th biggest of 100 data points should correspond to the z_{86} value of 1.08 since this is 86 percent of the way through the standard normal distribution. Therefore our 86th data value should roughly satisfy the equation

$$\frac{x_{86} - \bar{x}}{s} = 1.08.$$

This equation can be rearranged to give $x_{86} = s \times 1.08 + \bar{x}$. This manipulation can be performed for each data value to give $x_i \approx s \times z_i + \bar{x}$ or $x_i \approx \sigma \times z_i + \mu$. Since this is the equation of a straight line, plotting the values of x_i and the corresponding z_i should give an approximate straight line. This plot is known as a Q–Q plot. If the x_i do not come from a normal distribution, the Q–Q plot will tend to deviate from a straight line.

The P–P plot is based on the same principle, but the axes are marked in terms of proportions along the sample and the best matching normal distribution. In the P–P plot, the horizontal position of a data point gives the proportion of values up to the given point, whereas the vertical position gives the probability of obtaining less than the data value when selecting a number from a normal distribution with the same mean and standard deviation.

A further issue here is in the detail of how we should define how far along the distribution we are. We could be tempted to define the first or the last value of the distribution to be 0 percent or 100 percent of the way along, but 0 percent and 100 percent of the way along a normal distribution is $-\infty$ and $+\infty$, respectively. We see that defining starting or finishing percentages of the way through the sample distribution as 0 percent or 100 percent and then trying to get these to correspond to positions in a normal distribution won't be sensible. There is no best answer here. One method is based on the idea that in some sense a single value will "ideally" cut a distribution into two equal parts, two values will "ideally" cut a distribution into three equal parts, and so on. This principle applied to P–P and Q–Q plots is called Van der Warden's method.

In the illustration of P–P and Q–Q plots, the two plots on page 133 represent 100 values that were selected by a method that ensures that all values between 50 and 60 are equally likely to be selected. Values chosen this way are said to be values of a uniform random variable on the interval 50 to 60. For the 100 points from the uniform random variable on the interval 50 to 60, no observed probability accumulates until we get to just over 50, whereas the matching normal distribution "expects" some values here. The left-most point on the P–P plot therefore has a higher expected value than an observed value. Values in the very low 50s (more than two standard deviations away from the mean), however, are more crowded than are "expected" by the matching normal distribution. Observed probability here therefore accumulates more rapidly than expected probability and the line of points therefore increases in height relatively slow. Conversely, the normal distribution "expects" points around the mean of 55 to be relatively crowded compared to points from a uniform distribution and so the line of points here increases in height relatively rapidly. The situation for the points in the very high 50s mirrors the situation in the very low 50s. These considerations give the P–P plot its overall elongated "S" shape. Similar considerations explain the shape of the Q–Q plot. By comparison, the two plots on page 132 show values chosen at random from a normal distribution with the same mean and standard deviation as those from the uniform distribution. These values give plots almost on the ideal diagonal straight line.

END OPTIONAL

Statistical Tests for Normality

There are statistical tests (e.g., the Kolmogoroff–Smirnoff test or Lilliefors test) to see if it may be reasonable to believe that data come from a normal distribution. The underlying philosophy of these tests is similar to previous tests that we have covered. They do not answer the main question, "Do these data come from a normal distribution?" Instead, they answer the secondary question, "Could a data set coming from some normal distribution easily have the same single measure of its overall pattern as this data set?" If the answer to this secondary question is "yes," it is conventional to assume that it is reasonable to believe that the data do come from a normal distribution. Of course, in particular circumstances this convention may not actually be reasonable. The Komogoroff–Smirnoff test can be related pictorially to P–P plots. It is, roughly speaking, based on the greatest deviation of the P–P plot from the ideal straight line.

Using Common Sense

Finally, logic alone can tell us that certain data cannot be normally distributed. According to the normal distribution, values are centered around μ and values extending out more than one or two standard deviations on either side are unusual, but it is possible, though very unlikely, for values to occur many standard deviations on either side of μ. Indeed, it is possible for values to occur anywhere between $-\infty$ and $+\infty$. However, in many physical situations it is absolutely impossible for values to occur outside a certain range. For example, human height has to be a value greater than zero. If μ for human height was 170 cm and σ for human height was 10 cm, there would be probability of exactly zero of a person having a height smaller than 17σ below the mean, whereas the normal distribution says that there is a probability greater than zero (but exceedingly small) of such a height. In practice, height of adult humans of one gender could have a distribution so close to the normal distribution that there is no practical reason for assuming another distribution. However, this is not so true of human weight, where the mean may only be three or four standard deviations above zero: People 200 pounds above average weight are fairly unusual but not impossible, whereas people 200 pounds below average weight cannot occur. It would generally be inappropriate to assume that a normal distribution is a reasonable approximation when the mean is less than two standard deviations above zero and values below zero are physically impossible.

Procedure if Data Are Assumed to Be Normally Distributed

If the data consist of values of continuous variables and there is no good reason to think that the data do not at least approximately come from a normal distribution, statistical tests based on the normal distribution are generally used. The normal distribution is assumed for two reasons. It is at least approximately true that many sorts of measurements are approximately normally distributed. This is a consequence of the central limit theorem. The alternative to assuming a particular distribution, such as the normal distribution, is to use nonparametric tests of the sort discussed in Chapter 4, but this wastes a lot of information, using only the rank order of the data and not the actual values. Ideally, interpretation of the probabilities obtained from tests based on the normal distribution should then be tempered by some knowledge about the size of the inaccuracies that are likely to arise if, despite appearances, the underlying distribution is not normal.

Procedure If Data Are Assumed Not to Be Normally Distributed

If the data do not appear to come from an underlying normal distribution, there are two approaches to choose from. The first approach is to use nonparametric statistics. The second approach takes its inspiration from the observation that the lognormal distribution can be turned into a normal distribution by taking logs and from the observation that whereas the normal distribution or at least a close approximation to it seems to be the most common distribution in nature, the second most common distribution in nature is the lognormal distribution. The second approach is then to try taking the logs of the data to see if it might be reasonable to assume that the logs of the data values come from a normal distribution. If so, statistical tests based on the normal distribution are performed on the logs of the data values. This second approach can be taken further. If it is not reasonable to believe that either the original data or the logs of the data values have an underlying normal distribution, other mathematical manipulations of the data values are used. For example, we can look to see if square roots of all the data values could be reasonably regarded as normally distributed, do the same with reciprocals of each of the data values, or use some other mathematical manipulation of all the data values. This process is called *data transformation*. The aim is to find a transformation that makes it "look like" the transformed data are normally distributed. Once this is achieved, it is assumed that the transformed data are in fact normally distributed and statistical tests based on the normal distribution are applied to the transformed data. This approach may seem rather contrived, but nevertheless is quite commonly used.

SUMMARY OF STATISTICAL TESTS SO FAR

This section summarizes the situations in which the various statistical tests that we have described up until this point should be used. Most of the statistical tests covered so far are designed to help answer the question, "Is there a difference?" This question is asked in various contexts according to the source of the data and the different types of data:

Source of data	Dichotomous (e.g., better or worse) data	Numerical but not necessarily normal data	Numerical and normal data
Two related measures (e.g., measures before and after an intervention on the same individual, or measures on one twin who had an intervention and the other twin who didn't)	Sign test	Wilcoxon signed rank test	Paired samples t test
A single measure on two unrelated samples (e.g., measuring the same quantity on men and women)	Fisher's exact test	Mann–Whitney test	Independent samples t test

These tests are perhaps the main tests in basic statistics. However, in developing the theory of these tests we have also dealt with other tests. These include the following:

- The use of the binomial random variable to test an hypothesis about the proportion in the population.
- The use of the Poisson random variable to test an hypothesis about the value of the parameter λ or the average rate at which something happens.
- The use of the z test to test whether a single value or the average of a number of values comes from a normal distribution with μ and σ specified.
- The use of the single sample t test to test whether the average of a number of values comes from a normal distribution with just μ specified.
- The use of the Komogoroff–Smirnoff test to test whether data are normally distributed.

Once again, note that none of these tests directly answer the relevant question, "Is there a difference?" Instead, they answer the question, "Assuming there is no difference, how hard would it be for chance alone to explain the data?"

QUESTIONS

1. The demand for electricity in megawatts at peak load on any given day is said to be described by a normal random variable with expected value 90 megawatts and

standard deviation 10 megawatts (variance 100 megawatts squared). If this description of the demand for electricity is correct, what generating capacity must be available in order that the load be met 99.5 percent of the time?

2. All senior high school students in a certain state take a test on which the statewide results are approximately normally distributed with mean 60 and standard deviation 10. A random selection of sixty-four test results is chosen.

   a. What is the probability that the average of the sixty-four test results will be below 57?

   b. What is the probability that a school with sixty-four senior students will score an average result of less than 57?

3. Assume that in healthy American men the level of hemoglobin is normally distributed with mean $\mu = 14$ and standard deviation $\sigma = 1.1$.

   a. What is the probability that a healthy man chosen at random will have a hemoglobin of (i) *exactly* 14, (ii) between 14 and 15, or (iii) over 16.

   b. A laboratory wants to check the accuracy of its hemoglobin measurements. It therefore measures the hemoglobin of 400 healthy men, reasoning that the average of these measurements should be very close to the long-run average, which is known to be $\mu = 14$. If their method of measurement is accurate and the assumption that $\mu = 14$ and the standard deviation $\sigma = 1.1$ is also accurate, what is the probability that the average hemoglobin of 400 healthy men will be less than 13.9?

   c. Blood is obtained from a single randomly selected healthy man from Papua New Guinea (PNG). The hemoblobin level is 10.7. Do you now believe healthy men in PNG have lower hemoglobin levels?

4. Two different brands of racing bicycles are tested by five cyclists. The top speeds attained are given in the following list. Is there convincing evidence that one brand is superior?

	cyclist 1	cyclist 2	cyclist 3	cyclist 4	cyclist 5	cyclist 6
Brand A	30.5	41.2	29.8	35.9	25.7	40.3
Brand B	30.3	46.3	36.1	40.1	35.2	40.0

5. The time from purchase until major repairs are required is recorded for eight cheap bicycles and five expensive bicycles with the following results in days: cheap bikes—39, 117, 561, 57, 3, 27, 8, 2; expensive bikes—289, 641, 105, 111, 903.

   a. Is it reasonable to perform a parametric test (test based on a normal distribution) on these figures?

   b. Regardless of your answer to a, perform a parametric test; in addition, perform a nonparametric test.

   c. Do you believe that cheap bikes last as long as expensive bikes? Discuss.

NOTES

1. The terms and symbols here may be unfamiliar to those who have not completed high school mathematics. Exponentials and logs are ways of turning one number into

another and back again, just as squares and square roots turn one number into another and back again according to some rule. In the case of squares, the rule is "multiply the number by itself." In the case of exponential to base 10, the rule to convert, say, the number 5 would be to write down five "10s" in a row and put multiply signs between them. Exponential to base 10 of 5 is therefore $10 \times 10 \times 10 \times 10 \times 10$ or 10^5 or 100,000. Log to the base 10 is the reverse procedure, so it will turn the number 100,000 back into 5. For reasons that are beyond the scope of this book, it is often convenient to use a number e, which is approximately 2.718, in place of the simple number 10 in these procedures. When we simply write "exponential" and "log" and don't specify to base something, it is assumed that we mean to base e.

2. The expression "could easily" is taken as referring to a chance of more than 5 percent.

3. We could, however, use pairing, in that we could examine the length of sleep of men and women who form married couples. However, married couples may not be representative of the whole population of men and women. We could also match men and women who have been randomly selected so as to obtain pairs of men and women who are similar in age or some other attribute that might affect sleep, but we will assume that it is not convenient to do this.

4. More precisely, the F test is based on the figure f, which is the ratio of the sample variances of the two groups. If standard deviations are the same, f should be a figure near 1, but values away from 1 will occur by chance. The F test calculates the probability or p value for obtaining a value of f at least as far away from 1 as the observed value, assuming the underlying population variances of the two groups are in fact equal.

5. A special weighted form of averaging or pooling is used: s_p is obtained from a weighted average of the sample variances where the weightings used depend on the number of measures of spread in each of the samples. Other methods of estimating the amount of spread in the system could be used (e.g., the overall sample standard deviation could be calculated), but theory shows that it is only the use of s_p that will lead to values from a t distribution.

CHAPTER 6

General Issues
in Hypothesis Testing

The preceding chapters, as well as explaining the basis of the most common statistical tests, have strongly criticized the unthinking use of these tests in decision making. This chapter goes into more detail about general issues with hypothesis testing.

SELF-CONTRADICTORY IDEAS
UNDERLYING THE NULL HYPOTHESIS

Hypothesis testing starts with a null hypothesis: the hypothesis that the intervention has no effect on what is being measured. The figures are then examined to see if they provide evidence against the null hypothesis. This is the basis of an approach to scientific decision making that is entirely objective. It lets the figures alone provide the evidence. Recall that an objective approach to decision making was the fundamental aim of the inventors of statistics. However, the null hypothesis raises a set of ideas that are almost self-contradictory:

1. The null hypothesis is what we would prefer to believe in the absence of any evidence to the contrary.

2. The null hypothesis is something that we often hope is not true, for the way statistics tells us to "prove" that our new treatment works is to tell us to pretend that it doesn't work and then to see if the evidence talks us out of this null idea.

3. The null hypothesis is that the intervention has had absolutely zero effect, for apart from zero what other size effect would be natural to choose in any situation? Choose we must, for we use the null hypothesis to calculate a p value.

4. The null hypothesis is almost always a fiction. Almost all interventions will have some effect. The effect may be very small, or not worth having, but it will rarely be exactly zero.

Given these almost self-contradictory ideas involved in the null hypothesis, it seems odd to base scientific decision making on this hypothesis testing approach. The developers of statistics proposed a range of responses to cope with this criticism. Some of the responses are easy, perhaps too easy, and don't properly address the difficulties.

Responses to the Self-Contradictory Ideas Underlying the Null Hypothesis

Prudent Conservatism

The easy response to the near contradiction between ideas 1 and 2 is that hypothesis testing leads to a conservative approach to assessing the benefits of new treatment, the approach of "don't believe in anything new until you have to." Perhaps this is a wise approach, as there has been a history of overenthusiastic adoption of new treatments that haven't lived up to expectations and have sometimes turned out to be dangerous.

By good fortune there have been other benefits in the medical area from an unthinking blind adherence to the conservatism inherent in the use of null hypotheses. It has enabled ethics committees that have unthinkingly accepted the null hypothesis approach to declare as ethical some ethically doubtful clinical trials of new treatments and so has allowed these trials to proceed. This has led to more certain knowledge of the effectiveness of the new treatments. For example, say that there is a disease that is known to have a probability of ½ of killing someone using the standard treatment. A new treatment has been produced and it has been tried out on four people with the disease, all of whom survived. Now you find that you have the disease. Which treatment would you want? I think any person whose common sense hadn't been misled by an inappropriate statistics "education" would want the new treatment. However, those on the ethics committees have had some statistics "education." They will calculate the p value for the new treatment from the information that four out of four got better. If the new treatment was no better than the standard treatment, this result would occur $\frac{1}{2} \times \frac{1}{2} \times \frac{1}{2} \times \frac{1}{2} = 0.0625$ of the time, so this is the p value (assuming a one-tail test) (in the terminology of Chapter 4, we have found the probability that a binomial random variable Bi[4, ½] takes the value 4). Since 0.0625 is greater than 0.05, the ethics committee would declare that although it is clear the new treatment deserves further study, there is "no evidence" that the new treatment is worthwhile. Of course, the "no evidence" here should be replaced by "no convincing evidence," or even more appropriately by "some evidence, but not really convincing evidence." With this misunderstanding of the logic of statistics, it is then possible for the ethics committee to authorize a study on a larger number of humans with the disease, comparing some who get the new treatment with some who don't, without feeling guilty that those assigned to get the standard treatment may well be getting inferior treatment. The end result may be unfortunate for those patients in clinical trials who get the inferior treatment when there was already some evidence that it really was inferior treatment. On the other hand, there will be a long-term benefit in terms of the overall certainty of knowledge in medical practice.

However, there is another side to the story regarding the long-term benefits to medicine of statistical dogma. There are a huge variety of medical conditions and a large number of interventions that could be used with any of them. It would be almost impossible to do proper research studies and statistical analyses to check the effect of each possible intervention on each possible condition. Apart from the problem of the large number of studies required, there would be difficulties because of the amount of individual variability in the patients. This variability means that large numbers of patients would often need to be studied before interventions that have only small benefit stand out unequivocally as having an effect. Getting permission from ethics committees and permission from sufficient numbers of patients would be a problem, particularly in situations where it may be clear to both the ethics committees and the patients that common sense dictates that the intervention must be of some benefit. As a result, many interventions that common sense indicates almost certainly work have not been studied or have been studied using an insufficient number of patients. Since these interventions have not produced a p value less than 0.05, they are, at best, labelled as unproved. At worst it is declared that such interventions are of no value.

This is a huge practical problem in medicine. For almost every medical condition there will be a number of interventions that have been traditionally used and that common sense would indicate would be likely to be of at least some small benefit, but which have not resulted in a p value of less than 0.05 in research studies. Throughout medicine, "scholarly" education material for doctors contains simple assertions that such interventions are of no value. As a result, throughout medicine it becomes almost impossible to learn what an expert who used both common sense and a knowledge of the data from research studies would think was appropriate treatment.

A particularly dramatic example of the obsession with p values at the expense of common sense is the issue of hygiene in medicine. It has become commonplace for doctors to neglect cleaning the skin prior to injections. This neglect of basic hygiene is the result of the publication of recent statistical studies that have shown "no difference" in the infection rate between injections given into skin that has been cleaned and skin that hasn't been cleaned. However, we have known that cleaning prevents infection in other situations since 1848. In that year an obstetrician, Semmelweiss, convinced his colleagues that washing hands between the time of doing an autopsy on a woman who died of infection following childbirth and the time of delivering a baby from another woman prevented deaths from postchildbirth infection. Semmelweiss convinced his colleagues of this by cleaning his hands on a number of occasions between the postmortem room and the labor ward and deliberately not cleaning his hands on other occasions. At a considerable cost in women's lives, he thereby gathered enough evidence to convince his colleagues (this cost in lives is discussed in more detail later). Since Semmelweiss's time the discovery of germs has also provided a theoretical understanding of why cleaning ought to be useful in preventing infection in any situation. It should therefore be unnecessary to test the null hypothesis that in the situation of needle pricks, cleaning has no effect on the chance of infection. This really is a totally inappropriate null hypothesis, so the recent studies should not have been done. Those involved in such studies have divorced themselves totally from common sense.

Common sense since the nineteenth century had dictated that cleaning in any situation almost certainly helps prevent infection. In some situations it may make only slight differences. A slight increase in the rate of infection might not show up as "statistically significant," even in an experiment involving several thousand injections. Since infections are rare even following injections into uncleaned skin, there will be only a few infections in an experiment involving thousands and it may well be possible to blame the difference in the infection rate with and without cleaning on the sort of chance that occurs more often than 5 percent of the time. But infections can be extremely serious. Even a very small increase in the infection rate justifies the trivial costs of cleanliness. It is certainly not positive for knowledge in medicine that we should regard this common sense as unproved and of no value until a p value of less than 0.05 has been obtained in testing a null hypothesis.

The letters pages of the September 2001 issue of the *Medical Journal of Australia* (175: 341–342) gives an example of the breadth of misunderstanding regarding the use of statistics to make decisions about the value of clean injecting techniques.

Convenient Fiction

The near contradiction between ideas 3 and 4 concerning a null hypothesis—that the null hypothesis is the hypothesis of zero effect but an intervention will almost always have at least some tiny effect—can be brushed aside by saying that the null hypothesis is a convenient fiction. Almost all interventions may have at least some tiny effect, but in some cases the effect will be negligible and not worth knowing about and in other cases the effect will be important. We can hope that often an experiment will not find convincing evidence against the null hypothesis where the effect is negligible, but will find convincing evidence when the effect is important. How easy it is for an experiment to find convincing evidence that an intervention makes a difference depends on the background variability, the size of the difference made by the intervention, and the number of measurements. This is discussed in more detail later. For the moment, we note that our hope that our experiments will pick up large, important differences and overlook trivial unimportant differences will partly depend on the number of measurements made. Hopefully, most experiments involve numbers of measurements that will usually allow important differences to be detected but will usually miss unimportant differences. However, no general rule about the appropriate numbers of measurements is possible because the background variability and the size of the difference that we regard as important will affect the requirements. Therefore, there is no insistence in the philosophy of statistics about using appropriate numbers of measurements before arriving at conclusions about the null hypothesis.

We started with the criticism that the ideas behind a null hypothesis are almost self-contradictory, so starting with a null hypothesis seems an odd approach to scientific decision making. We have seen two easy responses to this criticism. The first response is that this approach to decision making leads to a conservative approach to assessing whether we should believe in the effectiveness of new interventions and that a conservative approach is often a good thing in the long run. The second response is that although the null hypothesis is a convenient fiction, experiments of the right size will often lead to useful conclusions about which effects really matter. However, we have seen that these responses are rather too easy. In the case of the first response we have seen that conservatism doesn't always match common sense, and in the case of the second response we have seen that there is no guarantee that our experiments will be of an appropriate size.

Don't Believe in Hypothesis Testing

The inventors of statistics prepared two more responses. The last of these responses is the topic of a following section: confidence intervals. The other response can be summed up as, "Don't believe in hypothesis testing!" Some elaboration is required here, of course. In hypothesis testing we start with the null hypothesis H_0 and see if there is sufficient evidence to talk us out of this null hypothesis. If there isn't, the correct conclusion, according to the advocates of statistics, is not to say "we believe in H_0"; instead, the form of words we should use is "we have insufficient evidence to reject H_0." If, on the other hand, we do have sufficient evidence to talk us out of the null hypothesis, evidence in the form of a p value of less than 0.05, we do state "we reject H_0." However, the founders of statistics were keen to point out that this is still just a provisional statement. An outcome from the experiment has occurred that, if H_0 were true, would be regarded as exceptional, something that occurs less than one in twenty times. However, something that occurs only one in twenty times will still sometimes occur, so our conclusion that this is evidence to reject H_0 should be regarded as only provisional. Unfortunately, the caution advocated here is often forgotten, and firm conclusions are instead drawn on the basis of whether p is greater or less than 0.05.

ERRORS

Since hypothesis testing consists of deciding between H_0 and H_a in the face of a background of variability and with a limited number of measurements, wrong decisions are inevitable. A more detailed discussion of these wrong decisions or errors is useful here.

There are two types of errors. A Type I error occurs when we conclude that H_a is true when really H_0 is true. A Type II error occurs when we conclude that H_0 is true when really H_a is true. The size of the Type I error is up to us, and conventionally we set the size of the Type I error as 0.05. This is equivalent to stating our benchmark p value is 0.05 (other terminology sometimes used is that the significance of the test is 0.05 or that the α level is 0.05). Why not set the Type I error to be zero? After all, we don't want to make any errors. A Type I error of 0.05 means that we will blame chance for any evidence that points toward the truth of H_a unless it is the sort of chance that occurs less than 0.05 of the time. If we set zero as our Type I error it would mean that we would blame chance for any evidence that points toward the truth of H_a unless it is the sort of chance that occurs less than zero of the time. In other words, no matter what the evidence, we would always believe in H_0. When H_0 happened to be true we would always get it right. However, when H_0 happened to be false the evidence that it was false would always be explained away by saying H_0 is true, but it just doesn't look that way because a series of coincidences came off. When H_a is true, we would always be making a Type II error. In the presence of uncertainty, we have to allow ourselves to sometimes be wrong when H_0 turns out to be true so as to allow the possibility that we can be right when H_0 turns out to be false. In general, the more we want to make a smaller Type I error (i.e., when there are no real long-run differences, we want to blame chance for all but the most extreme differences in our samples), the more likely we are to make a Type II error (i.e., when there are real differences, we will be more likely to wrongly believe that there is no real long-run difference and blame chance for the differences seen in our samples). There is a trade-off between the two types of errors.

The size we set for the Type I error is one of the factors determining the probability of a Type II error. There are three other factors: the background variability, the number of measurements, and the size of the difference made by the intervention. Why? Recall that a Type II error is made when H_a is true but starting from an initial assumption that H_0 is true we find insufficient evidence to talk us out of this incorrect belief. This can occur if there is a lot of background variability so that the evidence being obtained can easily be ascribed to chance effects. It can occur if only a few measurements are made so we can easily blame results in favor of H_a on a combination of just a few odd chances. Third, it can occur if H_a is not very different from H_0, so that results produced as a result of H_a being true will not be convincingly different from those that would have been produced if H_0 had been true. In general, the only factor we can control here is the number of measurements. Usually, the only way to lower both Type I and Type II errors simultaneously is to use more data.

Other terminology is sometimes used here. The size of the Type II error is sometimes given the symbol β, and $1 - \beta$ is called the "power" of the test. The power of the test is its chance of giving the answer that H_a is true when in fact H_a really is true. It follows that the power of the test is improved in the following situations:

1. If we are prepared to accept a larger Type I error (for example, we may set the benchmark p value to be larger than 0.05).
2. If there is less natural variability.
3. If there is a greater number of measurements.
4. If the effect of the intervention is large.

Ideally, experiments should be designed so that they have the power to generally detect differences that matter. On the other hand, for the sake of economy the numbers in the experiment should be limited so that the experiment need not have the power to usually detect differences that do not matter. Where the effect of the intervention is minimal, we will get results that make us reject H_0 hardly any more often than would have occurred if H_0 really had been true, since our true H_a is almost identical to our assumed H_0. In other words, the minimum power occurs when the intervention is of minimal effectiveness and this minimum power is the value of the Type I error or α. The power increases as the distance between H_0 and H_a increases until, when they are so far apart that they give unmistakably different results, the power is 1.

These definitions can be summarized in the following way:

	Decide H_0 is correct	Decide H_a is correct
H_0 truly correct	$(1 - \alpha)$	Type I error (α)
H_a truly correct	Type II error (β)	Power $(1 - \beta)$

There is a direct analogy here between hypothesis testing and medical testing. Specificity is the chance of a test telling us the person doesn't have the disease when in fact the person doesn't have the disease. In other words, it is the probability of the test indicating the correct decision when there is no disease. This is equivalent to the chance of not making a Type I error (i.e., specificity is equivalent to $1 - \alpha$). Sensitivity is the chance of a test telling us a person does have the disease when in fact it is true that the person does have the disease. Sensitivity is equivalent to the power of the test:

	Test indicates patient has no disease	Test indicates patient has the disease
Patient truly has no disease	Specificity	
Patient truly has the disease		Sensitivity

THE USE OF *P* VALUES AND COMMON SENSE IN HYPOTHESIS TESTING

The end result of all our statistical calculations to date has been a *p* value. Recall that our main question goes something like this: "Here are two groups. One group has had an intervention and the other group hasn't. The measurements on the two groups are not exactly the same. It looks like the intervention could have made a difference, but on the other hand the difference could be accounted for just by coincidence. What then is the cause of the difference? Is it for real, or does it just look that way because of coincidence?" The *p* value does not give us an answer to this main question. Instead, it gives an answer to the secondary question: "If we're going to blame coincidence for the difference, what sort of coincidence are we talking about?"

If we assume that it is coincidence alone that accounts for the difference and we then calculate that it is a very long coincidence that is needed to account for the difference between the two groups, it may be more reasonable to stop assuming that it is coincidence alone. Instead, we should now decide that the difference is there because the intervention makes a difference. A long coincidence here is traditionally taken to be the sort of coincidence that would occur less often than one in twenty times (i.e., $p \leq 0.05$).

However, it has been repeatedly emphasized that it is not always reasonable to use the traditional *p* value of 0.05 as a benchmark in deciding the answer to the main question. Sometimes the main question is stupid, as in the case of infection and cleaning the skin. We should never allow ourselves to blame chance for any difference in the infection rate after injections into cleaned skin compared to injections into uncleaned skin. Sometimes, while we may be prepared to believe in the null hypothesis initially, we may require only very weak evidence before giving it away. If I was only a bit skeptical about the ability of a relaxation tape to enhance sleep and I gave the tape to three people, all of whom said it improved their sleep the night they listened, then I would prefer to believe that the tape worked. Here, the *p* value is $\frac{1}{2} \times \frac{1}{2} \times \frac{1}{2} = 0.125$ (to calculate *p* we are assuming that H_0 is true—that the tape doesn't work at all—so it is a 50–50 chance [a probability of $\frac{1}{2}$] that each person will say better, and this chance has to come off three times in a row; hence, $\frac{1}{2} \times \frac{1}{2} \times \frac{1}{2}$). While without any evidence my best guess is that the tape doesn't work, I'm not strongly of that view. After seeing the result of a *p* value of 0.125, I think it would be more reasonable to believe that the tape works to improve sleep rather than thinking that it just seems that way because of a 0.125 chance coming off. On the other hand, sometimes I may believe strongly in H_0. Even if the next-door neighbor's child correctly guessed a number between 1 and 1,000 that I had written down in secret, I would rather believe that it was a fluke and not that the child is clairvoyant. Therefore, the benchmark *p* value for me to change my mind in this situation would be less than 0.001. We therefore see that one of the factors in choosing the benchmark *p* value (the *p* value that is just sufficient to convince a person about the alternative hypothesis) is our prior ideas about the relative chance of H_0 being correct versus H_a being correct.

Another factor we should consider is the consequences of being right or wrong. Recall that there are two ways we can be right and two ways we can be wrong, depending on whether H_0 or H_a is actually correct or incorrect (see the previous section). Now consider Semmelweiss and the washing of hands and imagine we are living in the mid-nineteenth century. If hand washing had really turned out to be of no benefit but in error (Type I) we decided it was of benefit, then we would be forcing doctors to wash their hands unnecessarily: This is a rather trivial cost. If, however, hand washing turned out, as we now know it has, to be of great benefit and we had decided in error (Type II) that it was not of benefit, then we would have incurred a great cost in human life. In the situation of Semmelweiss, the cost of making a Type II error was much greater than the cost of a Type I error.

Semmelweiss's colleagues should have taken this into account and should not have done as they did and forced Semmelweiss into prolonged human experimentation at considerable cost in human life before accepting that hand washing was worthwhile. We see from this example that in judging the evidence, as well as taking account of the *p* value and our prior ideas of the relative likelihood of H_0 and H_a, we should also take into account the costs of errors of both types. In general, we will do this in a subjective way. However, there is a branch of mathematics called decision theory that allows us to make the best decision on the basis of estimates of probabilities and costs and benefits.

CONFIDENCE INTERVALS

Hypothesis testing was the first approach of the founders of statistics to an objective method of decision making. However, we have seen that there are philosophical difficulties to this approach, in that the null hypothesis has to satisfy a near contradictory set of ideas. The various responses to these philosophical difficulties that we have discussed so far are not completely satisfactory. There are also serious practical problems in making decisions using hypothesis testing if *p* values are not properly understood and interpreted in light of a lot of common sense. A more comprehensive response to the philosophical problems of hypothesis testing is provided by the idea of *confidence intervals*. The theory and the calculations of confidence intervals follow directly from hypothesis testing, but the confidence interval approach does away with the null hypothesis and its attendant philosophical problems. Unfortunately, though, serious problems of interpretation remain if the confidence interval approach is used in making decisions without incorporating understanding and common sense.

To explain confidence intervals, consider the single sample *t* test. We will look at an example where we use this test in decisions concerning the average height of adult American women. We will compare the hypothesis testing approach and the confidence interval approach for the same data. For example, our hypothesis might be that the average height of American women is 170 cm and we may use the benchmark p value of 0.05 as our measure of how strongly we should hold on to this hypothesis. We note the arbitrary nature of both the hypothesis and the benchmark p value without further comment here. Let us say that we now choose two women at random from the population and find that their heights are 139 cm and 141 cm, so that the average height of our sample is 140 cm. We first look at the hypothesis testing approach. From Chapter 5 we know that under certain common circumstances, the formula

$$\frac{\bar{x} - \mu}{s/\sqrt{n}}$$

gives us a value chosen from a t distribution. In effect, this tells us to calculate the distance of the sample average from the hypothesized true average $(\bar{x} - \mu)$ and compare this to the estimated variability of our sample average

$$\frac{s}{\sqrt{n}}.$$

With two data values in our sample, theory tells us that the formula gives a value from a t_1 distribution. The estimated standard deviation of the numbers in this sample is

$$s = \sqrt{\frac{(139 - 140)^2 + (141 - 140)^2}{2 - 1}} = \sqrt{2},$$

so the estimated variability of our average (or estimated standard error or estimated standard deviation of the mean) is

$$\frac{s}{\sqrt{n}} = \frac{\sqrt{2}}{\sqrt{2}} = 1$$

Therefore, we can see that our sample average of 140 is 30 estimated standard errors from the hypothesized mean of 170. The t_1 distribution shows that 95 percent of the time the average of samples of two values will be within 12.71 estimated standard errors of the true mean. We would therefore reject the hypothesis that the true average is 170 cm using the $p = 0.05$ benchmark, since, if 170 cm was the true average height of women, then 95 percent of the time we would get an average for our sample of two in the range 170, give or take 12.71 estimated standard deviations of the mean, whereas our result of 140 is 30 estimated standard deviations of the mean away from 170. In other words, in light of the variability we see in the sample, our average of 140 is the sort of average that would "hardly ever" occur if the true mean was 170 and so we reject the hypothesis that the true mean is 170.

In the confidence interval approach we start with the same formula

$$\frac{\bar{x} - \mu}{s/\sqrt{n}}$$

and the information that 95 percent of the time this gives a value in the range −12.71 to +12.71. In our particular case, with

$$\frac{s}{\sqrt{n}} = 1,$$

we see that 95 percent of the time we would get a value of $\bar{x}$ such that $\bar{x} - \mu$ will be in the range −12.71 to +12.71. Since in our case $\bar{x} = 140$, we see that if μ was in the range 140 − 12.71 to 140 + 12.71 (or 127.29 to 152.71) then our value of 140 for $\bar{x}$ would not be unusual. It would not be unusual in the sense that if μ was in the range 127.29 to 152.71, the value 140 for $\bar{x}$ would be a value in the range of the more common values of $\bar{x}$ that can occur for such a μ. With one of these values for μ, our formula

$$\frac{\bar{x} - \mu}{s/\sqrt{n}}$$

gives a t value in the range −12.71 to +12.71, the sort of t values that occur 95 percent of the time. The range 127.29 to 152.71 is then referred to as a 95 percent confidence interval.

However, the terminology "confidence" is misleading. It almost implies "probability" without actually using the word. But probability it is not. The confidence interval calculation has used only part of the information that we would need to be truly confident about where μ might be. In the particular example here, one would have to have no concept of measurement or American women to believe that there is a 95 percent chance that the true average height of American women is in the range from 127 to 153 cm. Almost everyone knows that the true average height for American women is going to be a figure much closer to 170 cm. By sheer chance, our sample happened to give us a misleading idea of the average height of American women. Coincidence has led us to pick a sample of two dwarves.

The occurrence of such coincidences should not completely override our prior common sense ideas. Just because such a coincidence has occurred does not mean that we should now believe that there is a 95 percent chance that the average woman is between 127 and 153 cm tall. It would be more honest to refer to 95 percent confidence intervals as 5 percent compatibility intervals, as this terminology would emphasize that all that we have obtained is a range of possible values for μ that are compatible with the data. By compatibility, we mean that the data would not lead us to reject an hypothesis that μ was one of the values in the range if we used a benchmark *p* value of 0.05 or 5 percent. We need to take into account prior knowledge, if available, as well as compatibility with the data before we can make a sensible statement about a probable range for μ. Unfortunately, defining our prior knowledge in precise terms and incorporating this with the data is difficult. It also involves subjectivity: Different people are going to have slightly different prior ideas about likely values for the average height of women. Undertaking these tasks is the subject of Bayesian statistics. Conventional or frequentist statistics leaves us with confidence intervals. These should be understood as a range of values that is readily compatible with the data, but confidence intervals are commonly misunderstood as representing probability intervals.

Let us now look in more detail at the method for calculating confidence intervals in various situations. Consider our first and simplest test based on the normal distribution. This was the test of whether we should believe that Madagascans have the same height distribution as Americans based on the observation of one Madagascan. Recall that our Madagascan was 150 cm and that we were testing the hypothesis that the height of the Madagascan was a figure chosen at random from the distribution that describes the heights of the American population (i.e., N[170, 8²]). Our hypothesis testing approach asked whether the figure of 150 was the sort of figure we could "easily" get from this normal distribution. We answered this question by noting that 150 would come from a N(170, 8²) distribution as often as

$$\frac{150 - 170}{8} = -2.5$$

would come from a standard normal or **Z** distribution and then noting that we "hardly ever" (5% of the time) get figures from a **Z** distribution outside the range −1.96 to +1.96. The confidence interval approach here does not assume a particular value of 170 for μ, but we use the value 8 for σ. We then ask what values of μ would allow our figure of 150 to be "easily" obtained. Using the same manipulation as we used in hypothesis testing, this is equivalent to saying that we want μ so that

$$\frac{150 - \mu}{8}$$

is the sort of figure that is "easily" obtained from a **Z** distribution. The figures that are "easily" obtained from the **Z** distribution are figures in the range −1.96 to +1.96 in the sense that 95 percent of the time a figure chosen from a **Z** distribution will be in this range. Therefore, saying that we want μ so that

$$\frac{150 - \mu}{8}$$

is the sort of figure that is "easily" obtained from a **Z** distribution amounts to saying that we want μ so that

$$\frac{150 - \mu}{8}$$

is in the range -1.96 to $+1.96$. Mathematically, this can be written as

$$-1.96 < \frac{150 - \mu}{8} < 1.96.$$

To deal with this, note that if two things are unequal, then if the two things are multiplied by 8 they will still be unequal in the same way. This will also be true if 150 is subtracted from both of them. Therefore

$$-1.96 < \frac{150 - \mu}{8} < 1.96.$$

can be rewritten as $-1.96 \times 8 - 150 < -\mu < 1.96 \times 8 - 150$; that is, $-165.68 < -\mu < -134.32$. Now note that cancelling out minus signs from both sides of an inequality reverses the inequality: -4 is less than (on the negative side of) -3, but 4 is greater than 3. Using this principle, our statement $-165.68 < -\mu < -134.32$ becomes $134.32 < \mu < 165.68$. The range of values 134.32 to 165.68 is then our 95 percent confidence interval for the true long-run average height of Madagascans based on our observation of a single Madagascan and our assumption that the heights of all human populations are normally distributed with $\sigma = 8$. If we wanted 99 percent confidence intervals, in these calculations we would simply replace 1.96 by 2.576, as tables show that 99 percent of the time figures we get from a **Z** distribution are inside the range -2.576 to $+2.576$.

These calculations can be written as

$$P\left(-1.96 < \frac{X - \mu}{\sigma} < +1.96\right) = 0.95.$$

Therefore

$$P\left(X - 1.96\sigma < \mu < X + 1.96\sigma\right) = 0.95. \qquad (1)$$

If it is known that the value of **X** in our experiment is x, there is then the temptation to write

$$P\left(x - 1.96\sigma < \mu < x + 1.96\sigma\right) = 0.95. \qquad (2)$$

The range of values specified in the brackets in statement (2), $x - 1.96\sigma < \mu < x + 1.96\sigma$), is the 95 percent confidence interval for μ. However, statement (2) incorrectly states that there is a 95 percent chance that the value of μ is in this interval. Statement (2) does not follow from statement (1). Statement (1) is valid in telling us where values of X are likely to be knowing μ has a particular value. It is not valid to twist this around to statement (2) about where μ is likely to be given a particular value of X. The incorrect jump in logic here is analogous to a jump from a statement that "where there is smoke, there is a 95 percent chance of fire" to a statement "where there is fire, there is a 95 percent chance of smoke." In general, we cannot say that there is a 95 percent chance of μ being in the 95 percent confidence interval.

OPTIONAL

The reason the leap from statement (1) to statement (2) is not valid can be explained in other ways. Consider a number of drawers, each of which contains some red marbles and some black marbles. Let us say that it is known that in the first drawer 95 percent of the marbles are red. The statement, "Given that we are choosing marbles from drawer 1, there is a 95 percent chance the marble will be red," is equivalent to statement (1). Now let us say we opened an unknown drawer blindfolded and chose a marble and then found that it was red. The equivalent to statement (2) would be the statement, "Given that we have chosen a red marble there is a 95 percent chance that we have opened the first drawer." However, this statement would generally be wrong. The statement would certainly be wrong if there were lots of drawers to choose from and all but the first contained only red marbles.

Using the language and notation of Chapter 3, the difference between statement (1) and statement (2) is the same as the difference between the probability of A given B and the probability of B given A; that is, $P(A|B)$ and $P(B|A)$. The relationship between these two quantities was given in the development of Bayes's rule. It is

$$P(A|B) = P(B|A) \times \frac{P(A)}{P(B)}.$$

We will therefore have $P(A|B) = P(B|A)$ only if $P(A) = P(B)$. In this context, A is a statement about where μ is likely to be and B is a statement about where data values are likely to be. $P(A|B)$ is a statement about the probability of μ being in the 95 percent confidence interval knowing the data values. $P(B|A)$ is a statement about the chance of data values being in a certain range given a value of μ, and in this context its value is 0.95. Therefore, the statement $P(A|B) = P(B|A)$ will in effect be a statement that the 95 percent confidence interval has a 95 percent chance of containing the value of μ. We see from the earlier work that this last statement will only be true if $P(A) = P(B)$. In this context,

$P(A)$ will generally only equal $P(B)$ if the location of the data and the location of the parameter μ are both equally likely to be anywhere. If we have some prior common sense ideas about where these values are likely to be, it will not generally be true that the 95 percent confidence interval has a 95 percent chance of containing the value of μ. Instead, we should think of a 95 percent confidence interval as the range of values for the parameter μ that could "easily" have resulted in the observed data. Here, "easily" means that the data would not lead us to reject a hypothesis about the value of μ when our criterion for rejection is that the data are outside the range that would occur 95 percent of the time.

Unfortunately, most statistics texts do not go through an explanation like this to show why a 95 percent confidence interval does not necessarily have a 95 percent chance of containing the unknown parameter value μ. Explanation is omitted because it would involve statements about the likely location of the parameter μ. In the absence of data, this necessarily brings in statements about the location of μ that are not objective; that is, not based on data. The followers of frequentist statistics are so committed to objectivity that they won't use any explanation that brings in any hint of nonobjective ideas.

A practical example may make these ideas clearer. Say a chocolate manufacturer knows that he will be breaking the law and may get into trouble if the machinery that produces his 100-gm chocolate blocks is set to produce blocks that in the long-run average have a weight less than 100 gm. Let us say that he wants to check on his chance that he is breaking the law and so measures four chocolate blocks. These weigh 101, 105, 102, and 100 gm. This gives the mean $x = 102$ and $s = 2.16$. The theory discussed in Chapter 5 on the single sample t test tells us that

$$\frac{\bar{x} - \mu}{s/\sqrt{n}}$$

is a figure from a t_3 distribution. If the machinery was set so that he just failed to comply with the law, then μ might be 99.999 and

$$\frac{\bar{x} - \mu}{s/\sqrt{n}}$$

would then be

$$\frac{102 - 99.999}{2.16/\sqrt{4}} = 1.85$$

Figures as large as this from a t_3 distribution occur about 8 percent of the time. We could therefore say that if he was only just breaking the law he would get data that were misleading, in the sense of being at least as favorable as the data he got, 8 percent of the time. Similar calculations show that if he was breaking the law to the extent that his μ was 98.6 rather than 100, he could still get data at least as favorable as the data he got 2.5 percent of the time.

However, it is not logical to reverse these statements. The manufacturer can't look at these figures and say that there is an 8 percent chance that he is just breaking the law and a 2.5 percent chance that he is breaking it to the extent of having his machine set to 98.6. This would be illogical in the same way that the reasoning described earlier is illogical in the experiment in which a random drawer is selected and a random marble chosen. Knowing that 8 percent of the marbles in one particular drawer are red does not mean that if a red marble is obtained then there is an 8 percent chance that particular drawer has been chosen. More generally, the manufacturer could note that 95 percent of values from a t_3 distribution are between −3.182 and +3.182. The 95 percent confidence interval here is then given by the inequality

$$-3.182 < \frac{102 - \mu}{2.16 / \sqrt{4}} < +3.182 < +3.182, \text{ or } 98.6 < \mu < 105.4.$$

However, all this is answering the wrong question for the manufacturer. He wants to know, given the data, what the chances are that he is breaking the law. All that frequentist statistics can give him is the answer to the reverse question, "If he is breaking the law by a certain amount, what are his chances of getting the misleadingly favorable data he got?" Frequentist statistics does not allow him to jump from an answer to the reverse question to an answer to the original question. He could do so only if he had some prior idea about how badly he may have broken the law.

In particular, suppose that somehow the manufacturer, prior to measuring the sample, had reason to believe that there was a 90 percent chance that his machine was set to give an average block size of 99.999 ("av99999") and a 10 percent chance that it was set to give an average block size of 102.0 ("av102"). Maybe the manufacturer has just bought the chocolate factory from another businessman. Maybe there is a switch deep in the machinery and very hard to access that can be set so that the long-run average chocolate block size is either 99.999 or 102.0. The current manufacturer knows that the switch is there, though he is not yet able to check its setting, but he is 90 percent certain that the previous owner would have favored the illegal 99.999 setting. The manufacturer now also knows that the probability of a sample that averages at least 102 given the setting 99.999 is 0.08. In symbols we write P(s102|av99999) = 0.08. Using similar notation, we also know that P(s102|av102) = 0.5 (half the time the sample average will be at least as heavy as the set machine average). Bayes's theorem as discussed in Chapter 3, then gives us

$$P(\text{av99999}|\text{s102}) = \frac{P(\text{s102}|\text{av99999}) \times P(\text{av99999})}{P(\text{s102}|\text{av99999}) \times P(\text{av99999}) + P(\text{s102}|\text{av102}) \times P(\text{av102})}$$

so that

$$P(av99999|s102) = \frac{0.08 \times 0.9}{0.08 \times 0.9 + 0.5 \times 0.1} = 0.59.$$

Therefore, the manufacturer after measuring the sample should revise his prior belief that there was a 90 percent chance the machine is on the lower illegal setting. He should now believe that there is a 59 percent chance that the machine is on the lower illegal setting.

On the other hand, say that somehow the manufacturer knew that there was a 90 percent chance that his machine was set to give an average block size of 98.6 and a 10 percent chance that it was set to give an average block size of 102.0. Then, after obtaining the sample with average 102, the calculations can be repeated with $P(s102|av986) = 0.025$ in place of $P(s102|av99999) = 0.08$. The result will be that the manufacturer, after measuring the sample, should now believe there is a 31 percent chance that the machine is on the lower illegal setting. We see that the same data with the same confidence interval but combined with different prior ideas lead to different conclusions about probabilities.

In practice, the manufacturer's ideas about the machine setting prior to measuring the sample are usually going to be more complicated than a probability regarding a simple choice between the value 102.0 and one other value. The possible machine setting will be a continuous variable and the manufacturers prior ideas about likely settings would be represented by probabilities "smeared out" over a continuous range of values. Also note that the actual outcome is known not to be "102.0 or above," but "exactly 102.0." Taking all these factors into account, calculations like those here become very complicated and are the subject of Bayesian statistics. Even obtaining a reasonable mathematical representation of the manufacturer's prior ideas is a major challenge.

In frequentist statistics all that can be calculated is confidence intervals. As we have seen, it is not correct to think of the confidence interval as a probability that the manufacturer's μ is in the given range. There is no answer to the manufacturer's question in frequentist statistics. He cannot use the data to give him a probable range for μ. Statistics only answers the reverse question about how compatible the data are with various possible values of μ. In practice, though, he would be more confident that he was complying with the law if all of the 95 percent confidence interval for μ was above the legal value of 100.

END OPTIONAL

The single sample *t* test and *z* test examples are a bit different from most of our other tests in that they test an arbitrary hypothesis such as $\mu = 170$ rather than a null hypothesis H_0. Let us now look at how confidence intervals apply in situations where there is a null hypothesis. Consider the situation of two groups of people where one group has an intervention and the other group does not, and assume that we are measuring some outcome that is a value from a normal distribution. In the hypothesis testing approach we would look at the difference in the averages for the two groups and assess the variability in the system. Calculations would then allow us to answer the question, "Given this much variability in the system, how often would chance alone result in two group averages at least as different as these two group averages are?" The answer to this question is the *p* value and is the basis for our decision about whether the intervention makes a difference. The confidence interval approach instead works out that if the true long-run average difference as a result of the intervention was any figure out of some range of values then we could have easily obtained the average difference between the two groups that we have observed.

What do we mean by "we could have easily obtained"? To explain, assume that in place of the null hypothesis that there is no difference between the two groups we have a *d* hypothesis that there is a difference of some value *d* between the two groups. With a slight modification of the hypothesis testing calculations, we can test this *d* hypothesis. To test the *d* hypothesis, use the rule that we reject the *d* hypothesis if the actual difference in group averages are too far from the value *d*. By "too far" we mean that calculations show that if the *d* hypothesis were true then chance alone would lead to a group average difference at least as extreme as our observed difference less than 5 percent of the time. Such calculations would lead us to reject the *d* hypothesis where *d* is a long way from the observed average difference between the two groups, but we would not reject the *d* hypothesis when we chose *d* to be a value close to the observed difference in average between the two groups. There would be a range of values of *d* where we would not reject the *d* hypothesis. This range of values is the 95 percent confidence interval. If the true long-run average difference made by the intervention was any value in the 95 percent confidence interval, we could easily get the observed difference between the group averages in the sense that these observed values are not so far out from *d* as to make us think that we have convincing evidence against the value of *d*.

In place of a *p* value of 0.05 we could use any other *p* value as our rule for rejecting our *d* hypothesis. If we used a *p* value of 0.10, we would obtain 90 percent confidence intervals. If we used a *p* value of α, we would obtain $100 \times (1 - \alpha)$ percent confidence intervals.

Confidence intervals are a widely used extension to the analyses of paired sample and independent samples *t* tests. Recall that the paired sample *t* test is used when we make measurements before and after an intervention or when we treat one of a pair and not the other and compare measurements of outcomes. In the paired sample *t* test, under the null hypothesis the differences x_i between the items in each pair are figures drawn from a normal distribution centered on zero and with unknown standard deviation. This leads to the conclusion that

$$\frac{\bar{x}-0}{s/\sqrt{n}}$$

is a figure from a *t* distribution. With the confidence interval approach we would have a *d* hypothesis that the difference between the items in each pair is a figure drawn from a normal distribution centered on *d*. Then

$$\frac{\bar{x}-d}{s/\sqrt{n}}$$

would be a figure from a *t* distribution. If the computer or statistical tables told us that 95 percent of the time figures from the appropriate *t* distribution are between $-\tau$ and τ, then the 95 percent confidence interval for the difference *d* is given by the inequalities

$$-\tau < \frac{\bar{x}-d}{s/\sqrt{n}} < \tau$$

This can be rearranged to

$$\bar{x}-\tau\, s/\sqrt{n} < d < \bar{x}+\tau\, s/\sqrt{n}$$

Similarly, in the independent sample *t* test the null hypothesis leads to the result that the expression

$$\frac{\text{the difference between the two means}}{s_p\sqrt{\dfrac{1}{m}+\dfrac{1}{n}}}$$

will be a figure from an appropriate *t* distribution. Say one mean is $\bar{x}$ and the other is *y*; this expression can then be written as

$$\frac{\bar{x}-\bar{y}}{s_p\sqrt{\dfrac{1}{m}+\dfrac{1}{n}}}.$$

The equivalent *d* hypothesis is that

$$\frac{\bar{x}-\bar{y}-d}{s_p\sqrt{\dfrac{1}{m}+\dfrac{1}{n}}}$$

is a figure from a *t* distribution. This then leads to a confidence interval of the form

$$\bar{x} - \bar{y} - \tau s_p \sqrt{\frac{1}{m} + \frac{1}{n}} < d < \bar{x} - \bar{y} + \tau s_p \sqrt{\frac{1}{m} + \frac{1}{n}}$$

for the difference *d* made by the intervention. This rather messy expression is saying that the confidence interval for *d*, the true average difference between the groups, is within an amount

$$\tau s_p \sqrt{\frac{1}{m} + \frac{1}{n}}$$

of the observed average difference in our two samples, $\bar{x} - \bar{y}$. Different percentage confidence intervals are obtained by using different values for τ. Here τ is a figure from the appropriate *t* distribution such that values between $-\tau$ and $+\tau$ occur the required percentage of the time. These calculations are generally made by computer. Note that if the 95 percent confidence interval includes zero difference, then by definition we would not reject the null hypothesis that the difference is zero at the 5 percent significance level (i.e., $p > 0.05$).

For an example emphasizing the need to use common sense with interpretation of confidence intervals, consider the experiment mentioned in Chapter 1 on whether two half-hour music lessons per week are of more benefit than a single half-hour music lesson per week. In Chapter 1 we said that the experiment involved only small numbers and showed a surprisingly small 2 percent improvement in those who had more lessons. We will assume that the percentage units refer to marks out of 100 on an exam and it is reasonable to believe that these are approximately normally distributed (of course, values below 0% and above 100% are impossible, so the assumption of a normal distribution is at best approximate). We discussed how silly it would be to start with the null hypothesis, as it is not reasonable to believe, in the absence of evidence, that the extra tuition is of exactly zero benefit. If we looked at confidence intervals in some computer analysis of the figures, it might tell us that the 95 percent confidence interval for the amount of improvement was −18 to +22 percent. This would be telling us that if the true difference made by the tuition was anywhere between −18 and +22 percent we could "easily" get our figures. By "easily" we mean that our figures would not contradict a *d* hypothesis with *d* in this range when we used the benchmark *p* value of 0.05. The confidence interval would be wide like this, if there were very little data and what data there were indicated considerable variability. The confidence interval is symmetric around the observed mean of 2 percent improvement. This means that an actual improvement of zero (the null hypothesis) could give the observed figures just as easily as an actual improvement of 4 percent.

However, this does not mean that the null hypothesis that there is no improvement is just as likely as the hypothesis that there is a 4 percent improvement in marks of students who have an extra lesson each week. Common sense tells us that an extra lesson will almost certainly be of some positive value, not zero or negative value. The width of the confidence interval also tells us indirectly that our experiment was not very powerful. Even if the long-run improvement was as large as 22 percent, the confidence interval shows us that we could still "easily" have obtained our data that show only a 2 percent average improvement. If prior to the experiment I had been very uncertain about how much improvement could have been expected from an extra half-hour of tuition per week but I had guessed that it would probably make a difference on average of somewhere between 10 and 40 percent, then after seeing the confidence intervals given by the experiment I would reassess my guess. I might perhaps now guess that the true average long-run improvement from an extra half-hour per week of tuition would be a figure very likely to be in the range from 5 to 25 percent. Note that if the long-run average improvement was a figure bigger than 22 percent the confidence interval tells us it would be unusual to obtain data like ours with only 2 percent average improvement. However, this does not rule out the possibility that the long-run average improvement is more than 22 percent, and we must balance this against our feeling that 25 percent (the average of 10% and 40%) was our best guess prior to the experiment. On the other hand, I would continue to rule out the null hypothesis and negative values for the true long-run average improvement, even though the null hypothesis can give rise to our figure of a 2 percent average improvement more easily than a hypothesis that the true long-run average improvement is more than 5 percent. Common sense tells me that it is exceedingly unlikely that an extra half-hour of tuition can have no effect or a negative effect on performance.

Readers may be very unhappy with the amount of guesswork here. However, guesswork is unavoidable in attempting to make estimates in the face of limited and variable information. It is true that confidence intervals are based solely on calculation and do not require guesswork, but avoiding guesswork by stopping at a quote of the confidence intervals may not be a satisfactory solution. For example, 95 percent confidence intervals are widely misunderstood as being the range of numbers 95 percent certain to contain the true average value. If the situation allows any common-sense guessing or there is any prior knowledge, 95 percent confidence intervals cannot be thought of as being the range of numbers 95 percent certain to contain the true average value. The common sense and prior ideas have to be taken into account in deciding where the true average value is likely to be. This is a point that deserves emphasis.

Those who resent the speculation used in this example are, however, correct in some respects. We arbitrarily condensed our common sense into the guess that the extra tuition "probably" makes an average difference of somewhere between 10 and 40 percent. We then simply used another guess to combine our prior guess of 10 to 40 percent with our calculated 95 percent confidence interval of −18 to 22 percent to give our final guess of 5 to 25 percent as the range very likely to contain the unknown figure for true long-run average improvement. The proper method of assessing and combining the prior information with the information in the data is to use Bayesian statistics. Unfortunately, Bayesian statistics is more difficult and less amenable to automated calculation. It was briefly discussed in the chocolate manufacturer example in the optional material discussed earlier, but is not dealt with further here.

We can conclude by stating that confidence intervals are a useful extension to hypothesis testing. They sidestep the problem that the null hypothesis is almost certainly a fiction in most cases. In addition, statements about rejection or nonrejection of the null hypothesis give us no clue as to the sort of differences the experiment had the power to show, whereas the width of confidence intervals give us an idea of what sort of effects could be present and readily compatible with the data. However, just as in the case of hypothesis testing, confidence intervals must be interpreted with the assistance of common sense.

SUMMARY

- Hypothesis testing is objective, but it is not a logically satisfying approach to decision making.

- Hypothesis testing implies conservatism: not believing effects are real until the evidence is strong.

- Hypothesis testing will frequently fail to detect real differences when the differences are small, when numbers in the experiment are small, or when there is a lot of variability.

- The results of hypothesis testing should always be regarded as tentative.

- Hypothesis testing should always be interpreted in light of common sense and should take into account the chances of Type I and Type II errors and the costs of these errors.

- Confidence intervals give us a range of possible values for the long-run average difference caused by an intervention that are easily compatible with the data. They do not directly tell us the chance that the average difference is in some range. If we have no preconception about what difference is to be expected, the distinction between confidence intervals and probability intervals is not relevant. Otherwise, confidence intervals have to be interpreted with common sense.

In short, confidence intervals are a more satisfactory approach than hypothesis testing in decision making as they indicate the range of effects easily compatible with the data. However, confidence intervals also have to be interpreted in the light of common sense.

QUESTIONS

1. It is suspected that watching violence on television may have an adverse impact on human emotional development and social interaction. As part of a study on the issue, a number of young children are shown a nonviolent cartoon on one occa-

sion and a cartoon containing violence on another occasion. After each cartoon viewing, each child was introduced to a group of children of similar age (who watched neither cartoon) and their interaction with the group was studied. The number of times in which it was judged that a child socialized more constructively after the nonviolent cartoon than after the violent cartoon was assessed by an assessor who had no knowledge of the type of cartoon that had been shown.

a. What is the null hypothesis in this experiment? Is the null hypothesis appropriate? What is the alternative hypothesis in this experiment? Is a one-tail or two-tail test appropriate?

b. How plausible do you personally feel each of these hypotheses are?

c. In making a decision between these hypotheses, what do you see as the costs in making a Type I error? What do you see as the costs in making a Type II error?

d. What p value do you feel is appropriate to just make you change your mind between the two hypotheses?

e. Say that the experiment was actually performed on twelve children with the following results: Five children socialized equally well after both cartoons, six children socialized better after the nonviolent cartoon than after the violent cartoon, and one child socialized better after the violent cartoon than after the nonviolent cartoon. Perform an appropriate statistical test and find the p value. Do you now believe in H_0 or H_a?

2. In each of the following cases decide whether the hypothesis listed is appropriate as a null hypothesis that could be tested by collecting appropriate data. If you decide the hypothesis is not appropriate as a null hypothesis, state why not. If you believe that the null hypothesis is appropriate, state how strongly attached you are to the null hypothesis. How strongly attached you are to the null hypothesis should be expressed in terms of the rarity of event that would make you reject the null hypothesis. In other words, for certain moderate values of x you would not reject the null hypothesis if an event occurs that has 1 chance in x of occuring when the null hypothesis is true, even though the event is more likely to occur when the null hypothesis is false. The question is how big would x have to be before it was just enough to force you to change your mind and reject the null hypothesis. Give brief reasons for your opinions.

a. A rubber strip hanging between a car bumper and the road surface has no effect on motion sickness associated with car travel.

b. Short and tall people are equally good at basketball.

c. Male and female students are equally good at statistics.

d. The astrology column in a weekly magazine does not predict peoples' futures.

e. Vaccination against measles does not reduce your chance of catching measles.

f. Students learn just as well when lecture notes are written on a blackboard as when they are presented in the form of transparencies displayed with overhead projectors.

g. Motorcyclists and car drivers have equal chances of fatal road accidents.

h. New drug *A* is no more effective than aspirin at relieving a headache.

i. Gastroenteritis is as likely to spread in households where all members always wash their hands after using the toilet as it is in households where this is not the common practice.

j. Reticulated water is as good for plant growth as rainwater.

k. Small primary school classes have no educational advantage over large classes.

l. Drivers of four-cylinder cars have the same rate of fatal car accidents as drivers of six-cylinder cars.

m. There is no relationship between poverty and crime.

n. People who are blind in one eye are as good at distance perception as people who can see through both eyes.

o. Students from well-resourced government schools get as good an education as students from comparable private schools.

p. Beer stored in wooden kegs tastes the same as beer stored in metal kegs.

q. Forest regenerating after logging has as many different species of wildlife as old-growth forest.

r. Physically fit students have no academic advantage over unfit students.

s. The taste of equivalent amounts of sugar and saccharine in foods is indistinguishable.

t. Blackboards and whiteboards are equally popular among students.

u. Intensive fishing does not reduce the number of fish in the sea.

v. People who are deaf in one ear are as good as people with normal hearing in determining the direction of sound.

w. Brand *A* and brand *B* paints are equally long-lasting.

x. Numbers that have not appeared in the last 100 spins of the roulette wheel are no more likely to occur on the next spin.

y. Dalmatian dogs do not learn tricks more readily than golden retrievers.

z. Smokers and nonsmokers are equally likely to die prematurely.

aa. Men and women are equally likely to be victims of domestic violence.

bb. Stutterers are just as likely as those with fluent speech to gain employment in the public relations area.

cc. Toothbrushing does not protect against tooth decay.

dd. A medical history of heart attack does not affect longevity.

ee. Sanitary disposal of sewage does not necessarily reduce the incidence of diarrheal disease.

ff. Male and female cyclists are equally likely to have bike accidents requiring hospitalization.

gg. Marks on a statistics exam are not affected by additional tutoring.

hh. Marks on a statistics exam are not affected by an exercise program.

ii. Marks on a statistics exam are not affected by a carrot juice diet.

jj. Marks on a statistics exam are not affected by an ointment purchased from a traveling fair and rubbed into the forehead once per day for two weeks.

kk. Rich people and poor people are equally likely to steal food to eat.

ll. Rich people and poor people are equally likely to have surnames starting with the early letters of the alphabet.

mm. People who use sunscreen regularly are just as likely as people who don't use sunscreen to suffer from skin cancer later in life.

nn. Infants who display considerable enthusiasm for music are just as likely as other infants to show ability in mathematics in later life.

oo. Sagittarians have just average luck in terms of winning raffles.

pp. Clairvoyants who are consulted by strangers are not able to guess the birth dates of the strangers any more often than would be expected by chance.

qq. Consumption of fatty foods is not related to the later onset of rheumatoid arthritis.

rr. Possession of a beard is in men is not related to the later onset of rheumatoid arthritis.

ss. The day of the week on which you are born is not related to the later onset of rheumatoid arthritis.

tt. Family history of rheumatoid arthritis is not related to the later onset of rheumatoid arthritis.

uu. Potted plants exposed to the music of the Beatles grow just as well as potted plants exposed to the music of the Rolling Stones.

vv. A twenty-minute session of vigorous physical exercise in the late afternoon does not lead to improvement in sleep for people suffering from insomnia.

3. Consider each of the following null hypotheses and decide whether in testing each of these hypotheses you would use a one-tail test or a two-tail test. In each case give brief reasons for your choice.

a. Students who are provided with a private tutor visiting their home are as likely to get good results as students who are not provided with this service.

b. High school students whose main outside interest is sports are as likely to do well at school as students whose main outside interest is computer games.

c. Plants that are watered three times a week grow just as rapidly as plants that are watered daily.

d. Patients having major surgical procedures fare equally well regardless of whether the surgery is performed by general practitioner surgeons or fully trained specialist surgeons.

e. Cyclists wearing white clothes are just as likely to be hit by a car at night as cyclists wearing clothes of other colors.

f. People with blood group A are just as likely as people with blood group B to contract rheumatoid arthritis.

g. Blue-eyed and brown-eyed people are equally likely to suffer from rheumatoid arthritis.

h. Scuba divers who have been given a six-week training course are just as likely to have accidents as scuba divers who have been given a twelve-week training course.

i. Laboratory rats provided with unlimited quantities of wheat grain grow just as fast as rats provided with a variety of dinner table scraps in addition to unlimited quantities of wheat grain.

j. People whose surname starts with "A" are just as introverted as people whose surname starts with "Z."

k. Diabetics who have a weekly medical check suffer diabetic complications at the same rate as diabetics who have monthly medical checks.

4. A sample of chocolate bars from a factory are weighed and the results are 201.3, 201.7, 202.0, 202.4, 202.8, and 203.0 grams. Find

a. the 90 percent confidence interval for the population mean of the weight of a chocolate bar.

b. the 95 percent confidence interval for the population mean of the weight of a chocolate bar.

c. the 99 percent confidence interval for the population mean of the weight of a chocolate bar.

d. Is it unreasonable to believe, given these data, that the population mean of the weight of a chocolate bar is 200 grams?

CHAPTER 7

Causality: Interventions and Observational Studies

Up until now we have often referred to interventions. We have often talked about comparing two groups. One has been a comparison or control group. The other group has been subjected to some intervention. In the first example in this book the intervention was an additional half-hour of music tuition per week, but we could imagine an endless variety of interventions that we might want to test in various circumstances. In such cases, provided the subjects for both groups are chosen at random from the same population, there can be only two explanations for any differences between the two groups: chance and the effects of the intervention. Because there is variability, the response of the subjects selected for the control group may just by coincidence be different enough from the subjects selected for the intervention group to make the average of measurements of the two groups appreciably different from each other even though the intervention itself was entirely useless. On the other hand, the average of measurements of the two groups would be likely to be different if the intervention has a real effect. If we denote the intervention by the symbol A and the difference by the symbol B, then we can summarize these two possible explanations as

- chance
- A implies B (in symbols we write A $\Rightarrow$ B).

Often, however, the comparisons that we want to make are between two groups that are different by nature. For example, we might want to compare measurements of the lung function of drinkers and nondrinkers of alcohol. The difference between the two groups here is in whether the individuals drink. In a sense, this is the intervention. However, we did not create this intervention. We just observe its effect, so this is called an observational study. An intervention is often not ethical or feasible. Clearly we could not collect a large number of people who didn't care either way about whether they took up drinking, tell half of them to start drinking and the other half not to, and then later measure their lung function. Instead, we choose two groups of people who have made the different decisions for themselves about drinking. The two groups may then be consistently different from each other in more than just their drinking habits. In that sense, the two groups do not come from the same population. This has important implications for causality. Say we find poorer average lung function measurements in the drinkers compared to the nondrinkers. There are now four possible explanations:

- chance.
- drinking (A) causes poorer lung function (B).
- poorer lung function (B) causes drinking (A).
- some third factor (C) causes poorer lung function (B) and separately causes people to drink (A).

In symbols, the four possible explanations are

- chance.
- $A \Rightarrow B$.
- $B \Rightarrow A$.
- $C \Rightarrow$ both A and B.

The last possibility, where some third factor causes poorer lung function and separately causes people to drink, requires more explanation. Perhaps a smoking habit tends to induce people to go to bars more frequently to purchase cigarettes. Consequently, these people then tend to become drinkers. The smoking also directly affects their lung function. Hence, even though drinking may do nothing to cause lung damage directly, drinking and lung damage will be associated because smoking has tended to cause both drinking and lung damage. There is a more realistic but less direct version of this scenario that will give a similar result. Certain social and personality factors result in some people being more concerned with immediate gratification than with long-term consequences. Many such people will tend to become both smokers and drinkers, so smoking and drinking are associated. We would again have an association between poorer lung function and alcohol because of the third factor, smoking.

It should be noted that we have used the word "intervention" without comment when we have been dealing with observational studies in a few of the previous examples. The statistical analysis is the same regardless of whether the intervention is one we create in an experiment or one we observe. However, in an observational study, when we decide that the differences between two groups are too big to reasonably be put down to chance we have to consider the options B $\Rightarrow$ A or C $\Rightarrow$ both A and B, as well as the most obvious possibility that A $\Rightarrow$ B. Common sense, not statistics, can help sort out these options.

Often we can rule out one of A $\Rightarrow$ B or B $\Rightarrow$ A by looking at which came first in time. Reasonable people believe that smoking is a cause of lung cancer. The smoking habit usually precedes the lung cancer by many years. A few diehard smokers have argued that an intuitive knowledge that one is doomed to lung cancer induces people to take up smoking, a habit that they argue protects the lungs from cancer. They continue their argument by stating that although smoking is protective to some extent, it is only partly successful in preventing the otherwise inevitable lung cancer: hence the association between smoking and lung cancer. There are at least three findings that do not fit this explanation: Dogs attached to smoking machines develop lung cancer, there are known cancer-causing chemicals in cigarette smoke, and there is a proportionality between lung cancer incidence and prevalence of smoking in populations in various parts of the world and at various times. However, even if a person knew nothing about any of these findings, the arguments of the diehard smokers would still seem far-fetched. The argument seems far-fetched because what is to happen in the future, the lung cancer, must affect the decision on whether to take up smoking in the past. This then requires the invention of a possible mechanism to explain how this time reversal could happen. Rather than invent such mechanisms, it will generally be more satisfactory to simply concede that A and B are associated for reasons other than B $\Rightarrow$ A.

It is also relevant to note that the word "cause" is a loose word that has the precise meaning of either "contributory cause" or "absolute cause." Smoking is a contributory cause of lung cancer. Not everyone who smokes gets lung cancer, but smoking increases the chance that a person will get lung cancer. There are other factors involved in lung cancer, such as exposure to asbestos, radioactive particles, industrial pollution, aging, and unknown factors that can be lumped together as "bad luck." On the other hand, destroying a forest is an absolute cause of the death of animals in the forest, assuming it is known that these animals are totally unable to survive without the habitat provided by the forest. A person who objects to the bland statement that "smoking causes lung cancer" on the grounds that not everyone who smokes automatically gets lung cancer is interpreting "cause" as "absolute cause," whereas most people would gather from the context that "contributory cause" is meant. In statistics we are generally interested in contributory causes. Absolute causes usually cannot be confused with chance effects; therefore it is not necessary to use statistics to assess them.

The problem of eliminating the possibility that C $\Rightarrow$ both A and B is much more difficult. For example, it is known from observational studies that older women who take hormone replacement therapy (HRT) have fewer heart attacks than women who don't. However, at the time of writing there is considerable controversy between those who conclude that A $\Rightarrow$ B (i.e., HRT causes a decrease in heart attacks) and those who argue C $\Rightarrow$ both A and B. Those who argue C $\Rightarrow$ both A and B believe some other factor causes some women to take HRT and separately causes a decrease in heart attacks. The other factor could well be attention to health. Women who are focused on their health are more likely to attend doctors for the distress caused by menopausal symptoms and will then receive HRT, and are also more likely to avoid heart attacks by living a generally healthier life. It is necessary to await studies in which women are assigned at random to a HRT group and a non-HRT group instead of allowing women to self-select into these groups in order to be certain about causality here.[1]

DOUBLE BLIND PLACEBO-CONTROLLED TRIALS

In the medical setting, to decide between the options of A $\Rightarrow$ B, or C $\Rightarrow$ both A and B experiments are preferable to observational studies where patients self-select. In an experiment, unlike a study where patients self-select their treatment, there should be no difference between patients assigned to various treatments other than that due to random chance. However, further care is often needed before an experiment on humans can properly assess the benefit of any new treatment. The ideal form for such experiments is known as a double blind placebo-controlled clinical trial. A placebo is a dummy treatment. If patients in a study are aware that they are getting a new treatment that may turn out to be more effective than basic care, the power of positive thinking may affect the results. The power of positive thinking is often enough to give those who think that they have received the new treatment a better outcome than those who think they have received no treatment. Positive thinking is known to have some effectiveness in a wide range of medical conditions, from perception of pain and depression to skin rashes and warts. To ensure that the power of positive thinking works equally for people in both groups, a placebo is given to one group and both groups are kept ignorant of whether they are receiving the new treatment or the placebo. So powerful and contagious is the power of positive thinking that it is necessary to prevent the doctor administering the treatments from knowing which is the new treatment and which is the dummy treatment or placebo.

Since both patient and doctor are blind to the nature of the treatment, we have the name "double blind placebo-controlled trial." The term "controlled" here implies the existence of a comparison group. Double blind placebo-controlled trials are easy to arrange if the treatment consists of a drug that can be taken as a tasteless coated tablet and has no immediate side effects that would make it clear to the patient that they were receiving active treatment. The placebo can then be a similar tasteless coated tablet containing only sugar. A double blind placebo-controlled trial with such sugar tablets would be used for the assessment of a new headache tablet, for example. However, double blind placebo-controlled trials may be difficult or impossible in the case of treatments such as acupuncture or surgery. A double blind placebo-controlled trial in the case of surgery would need to involve a surgeon not in communication with the patient or the experimenter. That surgeon would then make a cut through the anaesthetized patient's skin but not perform the internal surgery. Such studies have actually been done, but there are obvious ethical difficulties: The placebo is not harmless like a sugar pill; it involves an anaesthetic and an incision. Double blind placebo-controlled surgery is a rarity. Instead, the placebo effect is usually ignored in the case of surgery and experiments of new surgical approaches consist of "trials" in which one group receives the new surgery and the other group, the "controls," receives the old standard care.

PROSPECTIVE AND RETROSPECTIVE STUDIES

There are many situations other than surgery in which double blind placebo-controlled trials are not possible. Often we want to assess the effect of differences in lifestyles where patients have self-selected into two major groups. Usually it will not be feasible for the experimenter to impose these differences on two groups chosen from a single population. For example, it is not feasible to use a controlled trial to assess the effects on health of lifetime dietary preferences. In such situations we must use observational studies. There are two major sorts of observational study in the medical context: prospective studies and retrospective studies.

In a prospective study, people who have self-selected into two groups are followed over a period of time to see if they develop a particular condition. For example, smokers and nonsmokers have been followed over a long time to study the incidence of lung cancer in both groups. In such a prospective study we know for certain that the smoking precedes the lung cancer, so if we find that there is an association between smoking and cancer that is too strong to be reasonably put down to coincidence we can virtually eliminate the possibility that lung cancer causes smoking because the smoking came first. However, other information (such as knowing that smoke contains cancer-causing chemicals) is needed to eliminate the possibility that $C \Rightarrow$ both A and B and so conclude that $A \Rightarrow B$ (i.e., smoking causes lung cancer). Prospective studies are expensive, particularly for less common diseases, for they involve following a large number of people for a long time until sufficient numbers of cases of the disease develop.

In a retrospective study, those who already have contracted a disease are asked about their previous exposure to a possible risk factor and their answers are compared to answers from a control group who don't have the disease. Although retrospective studies are cheaper than prospective studies, there are more pitfalls. We can no longer ignore the possibility that any association is due to $B \Rightarrow A$. Those who are victims of a disease may be more likely to recall exposure to possible risk factors, so by selective recall the disease can in effect cause the preceding risk factor. In addition, there are major problems in selecting the control group for comparison. In the case of lung cancer and smoking, if a control group was selected at random from the total population the most striking difference between the two groups would probably be age. The average age of a sample from the total population is likely to be in the twenties, whereas the average age of lung cancer cases is likely to be several decades older. This difference in age would not be telling us anything we did not know before, but on the other hand, the difference in age could account for many differences between the two groups that had nothing to do with the real risk factors for lung cancer. To overcome this problem, some matching by age and perhaps other factors is necessary. In other words, the control group should be selected so that they are similar in age and perhaps similar in some other ways to the lung cancer group. However, overmatching is then a problem. If a study matched lung cancer sufferers and non–lung cancer sufferers according to smoking habit, the study would have excluded the possibility of assessing the most important risk factor for lung cancer: smoking.

There is a further problem. In practice, control groups are likely to consist of people who have the time and generosity to cooperate with a medical study even though they will not benefit personally. They are therefore likely to have, on average, different personalities from the general population and from sufferers. Differences in exposure to various factors between the control group and the sufferers may therefore reflect the differences in personality and not be due to the factors being risk factors for the disease. Ideally, the control group should be chosen so that if one of the sufferers had somehow not caught the disease that person would have the same chance of being included in the control group as others in the population from which the control group is selected. This is a difficult criterion to specify or even understand properly. It is almost impossible to put into practice. Hence, retrospective studies are less reliable than prospective studies.

SUMMARY

- In an experimental study two groups of individuals are chosen from the same population. One group is subjected to an intervention and the other group is kept as a control group for comparison.

- In an experimental study there are only two possible explanations for differences between the two groups: chance or the effect of the intervention (chance or A$\Rightarrow$B).

- In medical studies it is desirable to distinguish between the effect of an intervention that is due to the power of positive thinking and the underlying physiological effect of the intervention. This is achieved by an experimental format known as the double blind placebo-controlled clinical trial, where neither experimenter nor patient are aware of who is getting the active treatment and who is getting the dummy.

- In an observational study people self-select into two groups, with one group exposed to some factor of interest and the other group not exposed. The two groups will generally be different from each other in other ways as well. In an observational study there are four possible explanations for differences between the two groups: chance, A ⟹ B, B ⟹ A, or C ⟹ both A and B.

- Observational studies are of two main types. Prospective studies are more expensive but are more reliable in establishing the direction of causality. Retrospective studies are cheaper but are less reliable.

QUESTIONS

1. Think of an example where all three possible directions of causality—A ⟹ B, B ⟹ A, and C ⟹ both A and B—are likely to be operating simultaneously.

2. A retrospective study is set up to examine the risk factors for suffering a broken right leg, A control group is obtained that consists of those who have suffered a broken left leg. Would this study reveal that motor bike accidents are the main cause of suffering a broken right leg?

3. A prospective study of heart attacks in a certain city enrolled 250 randomly selected men aged fifty to fifty-five with no history of heart trouble. In other words, these men were identified as living in the city in which the study was carried out and a check was made that they fell into the required age group and had not had heart trouble. It was noted that 200 of the 250 men frequently attended football matches as part of their Saturday entertainment, whereas the 50 remaining men preferred classical music concerts for Saturday entertainment and rarely attended football matches. After ten years it was found that forty of the football goers had suffered heart attacks, whereas only three of the concert goers had suffered heart attacks. Fisher's exact test (two-tail) performed on these data show that $p = 0.020$. Do you believe that this study provides reasonably convincing evidence that men who prefer classical music to football are less likely to suffer heart attacks? Do you believe that this study provides reasonably convincing evidence that attending football matches tends to increase a man's chance of having a heart attack? Discuss.

4. A sample of 2,000 people had their serum cholesterol level measured and were then followed up for five years. At the end of the five years it was found that 20 of the 2,000 had met a violent death. Seventeen of the twenty were those who had a below median level of cholesterol. Discuss the p value you would use to make a decision between the hypotheses here. Perform an appropriate statistical test. Do you believe that a low cholesterol predisposes people to violent death? Would your decision be changed if you were told that 200 of the original 2,000 had died of a range of causes and statistical tests looking for an association on each cause of death had been performed separately? Would your decision be changed if you were told that the study had been conducted over thirty years rather than five years and that by the time the results were analyzed, in addition to the 20 who had died violent deaths, 1,800 had died of other causes?

5. A retrospective study is set up to examine the risk factors for parenting a mentally retarded child. The parents of all children with mental retardation in a given region all agree to cooperate in answering an extensive questionaire covering a large range of factors that could possibly be relevant to mental retardation. Neigh-

bors of the families with retarded children are given the same questionaire. If the nearest neighbor is not prepared to cooperate and/or has no children, the questionaire is given to the nearest neighbor who will cooperate and has children. Would this study be capable of detecting the following possible causes of mental retardation?

a. Mother suffering from rubella (German measles) in pregnancy.

b. Parents having many children.

c. Age of parents.

d. Mother suffering from poverty and malnutrition.

e. Residence near a lead smelter.

NOTE

1. As this book goes to press, the result of the HERB randomized controlled trial on HRT has just been released. Surprisingly, it shows that the indidence of heart attacks in women on HRT is greater than the indidence in women not on HRT. If this finding is "for real" and not the result of chance, it tells us that a real but small negative effect of HRT on health was so overwhelmend by the tendency discussed here, where "C $\Rightarrow$ both A and B," that doctors had gained the opposite impression from observational studies.

CHAPTER 8

Categorical Measurements on Two or More Groups

The very first statistical test discussed in this book was Fisher's exact test. This test applies when we have two groups and we can categorize the outcomes in these two groups in two ways. For example, the groups could be "fired workers" and "not fired workers" and the categorization could be "female" and "male." Since this was our first test, we did not have the background to fully explain a number of features of this test. We will do so now before studying other tests where the outcome is a category rather than a numerical measurement.

MORE ON FISHER'S EXACT TEST

We first note that a table displaying the results is called a 2 × 2 contingency table (the 2s here refer to the fact that—excluding the headings and totals—there are two rows and two columns. The table contains 2 × 2 = 4 "boxes" of data). We also note that the data can be viewed in two equivalent ways. In our first example, the groups were "fired" and "not fired" and the categories were "female" and "male," but the groupings and categorization can be swapped around. The two groups could be "female" and "male" and the categories could be "fired" and "not fired." There is a symmetry here. Most methods of analyzing data are not affected by swapping the definition of groups and categories.

In Chapter 3 we explained how the contingency table on page 176 led to the p value calculation

$$p = \frac{4}{9} \times \frac{3}{8} \times \frac{2}{7} \times \frac{1}{6} \approx 0.008 < 0.05.$$

This calculation was based on the reasoning that with nine people, four of them men, we had a 4/9 chance of obtaining a man for our first butter preferrer. Then, with three men and eight people remaining, we had a 3/8 chance of obtaining a man for our second butter preferrer, and so on. We did not deal with contingency tables where there are all non-zero entries in the table. The calculations then become complicated, involving more extensive use of the theory of combinations (see Chapter 3).

	Male	Female	Totals
Prefer butter	4	0	4
Prefer marg	0	5	5
Totals	4	5	9

For those who want it, the theory is given here, though it is not essential for our purposes to go through the details. While the theory is not essential, the discussion afterward is necessary for a proper understanding of the p value when the 2 × 2 contingency table doesn't have zeros on the diagonal.

OPTIONAL

For those who are interested in the theory, given the accompanying table we reason that there are $^{a+b+c+d}C_{a+c}$ ways of choosing $a + c$ people out of the total of $a + b + c + d$ people to be in Category A.

	Group I	Group II	Totals
Category A	a	c	$a+c$
Category B	b	d	$b+d$
Totals	$a+b$	$c+d$	$a+b+c+d$

The number of ways of choosing a Group I people to be in Category A and c Group II people to be in Category A is $^{a+b}C_a \times {}^{c+d}C_c$. We compare this number of ways of obtaining a in the top left box of the 2 × 2 contingency table and c in the top right box of the table with the $^{a+b+c+d}C_{a+c}$ ways of obtaining the total $a + c$ for the sum of the top two boxes in the table. The probability that the $a + c$ in Category A will be made up of this particular choice from the two groups is then

$$\frac{^{a+b}C_a \times {}^{c+d}C_c}{^{a+b+c+d}C_{a+c}}$$

With the totals fixed, once we have fixed the number a in the top left box of the 2×2 contingency table the numbers in the remaining three boxes are fixed also. The expression

$$\frac{^{a+b}C_a \times {}^{c+d}C_c}{^{a+b+c+d}C_{a+c}}$$

is therefore the probability of this particular allocation of data to the four boxes of the 2×2 contingency table when the totals are as specified.

Those who have made the effort to have understood the logic in the previous paragraph will now be disappointed to learn that there is a subtle philosophical objection to Fisher's exact test. The question that is usually most appropriate is, "What is the chance of getting a Group I people in Category A and c Group II people in Category A given that the chance of a person of either group being in Category A is unknown?" whereas the question that we have answered with the Fisher's exact test calculation is, "What is the chance of getting a Group I people in Category A and c Group II people in Category A given that $a + c$ people out of the entire sample are in Category A?" For either question we want an answer based on the null hypothesis assumption that there is no preference for those in a particular group to tend toward a particular category. The latter question, which we have answered exactly, would be completely appropriate only if we had selected $a + b$ Group I people and $c + d$ Group II people, all of whom were initially in just one of the categories. We then waited for a disease or some other process over time to change people's categories until we knew that there were exactly $a + c$ people out of the entire sample in Category A. Usually, though, we do not predetermine the total numbers in the categories and so the question that we have answered is not quite the appropriate question. We cannot obtain an exact answer to the more appropriate question, "What is the chance of getting a Group I people in Category A and c Group II people in Category A given that the chance of a person being in Category A is unknown (but not affected by which group the person is in)?" An exact answer is unobtainable because it is based on the unknown chance that someone will be in category A, and we cannot do calculations when the underlying chance for an individual is unknown. However, when numbers are large the proportion in the two groups combined that we actually see in a category is a good guide to the underlying chance of an individual of either group being in a category. Some rather difficult theory based partly on this idea allows an approximate answer to the more appropriate question when numbers are large. The approximate answer is in the form of a statistical test called the chi-square test, which is discussed later in this chapter.

END OPTIONAL

We now return to the issue of what the p value actually measures when we are dealing with a 2×2 contingency table that doesn't have zeros on the diagonal.

Say that we suspected that there might be a tendency for females to have different butter–margarine preferences compared to men, and instead of the results in the first table of the chapter we obtained the accompanying results.

	Male	Female	Totals
Prefer butter	5	2	7
Prefer marg	1	5	6
Totals	6	7	13

We will assume that the null hypothesis is reasonable here. In other words, in the absence of data we are happy to take as our starting point that there are no gender differences between men and women in terms of butter–margarine preferences. We want to ask the primary question, "Given the results in the accompanying table, how likely is it that there is a gender difference between men and women in terms of butter–margarine preferences?" As always, statistics doesn't directly answer this question but instead answers a secondary question. This secondary question is something like, "If there was no gender difference in butter–margarine preferences, how often would pure coincidence lead to the results here that make it look like there is a gender difference in butter–margarine preferences?"

However, the secondary question is actually one step more complicated than this. As explained later (and previously on pages 67 and 68), it is necessary to not just ask about how likely our particular results are under H_0. Instead, we need to ask about the combined likelihood of all the outcomes that are at least as extreme in favor of H_a as our particular result. In doing so, we are thinking that there are a range of possible outcomes that would be most likely if H_0 were correct, and outcomes in this range commonly occur. We are content to continue to believe in H_0 if we get one of these results. Outcomes outside this range "hardly ever" occur and make us reject H_0. We define "commonly" occur and "hardly ever" occur by a benchmark p value that we will denote p_b. Outcomes that occur "commonly" are part of a range of outcomes that occur $1 - p_b$ of the time, whereas the remaining outcomes that occur "hardly ever" are part of a range that occurs p_b of the time. When we get a particular outcome from an experiment, we ask what the combined probability is of all the outcomes that are at least as extreme as this outcome: The answer will tell us whether we are dealing with the sort of outcome that occurs "commonly" or "hardly ever," assuming H_0 is true. In the context of Fisher's exact test, recall that under H_0 we are assuming that it is a given fact that we have selected a certain number of men and women and a certain number of butter preferrers and margarine preferrers and that chance then allocates the numbers in the various categories. Overall, about half of the people prefer butter and about half are women, so if H_0 were true overall we would expect about half the women and half the men to be butter preferrers. We would therefore expect about three men and three women to be butter preferrers, but we see that we have fewer women and more men than we would expect. The set of outcomes that are at least as extreme as those we have obtained are, the outcome we have actually obtained and all possible outcomes with even fewer women and even more men in the butter-preferrer category. In working out what these outcomes are, we must remember that we are taking it as a given that we have selected six men, seven women, seven butter preferrers, and six margarine preferrers, and that the allocation of the men and women into the butter or margarine category has occurred purely by chance. The table giving a more extreme possible outcome is given here.

	Male	Female	Totals
Prefer butter	6	1	7
Prefer marg	0	6	6
Totals	6	7	13

More extreme outcomes than this are not possible, as we are assuming the numbers in the categories are fixed and we have used up all our males in the butter-preferrer category. Calculation shows that the probability of the outcome given in the table on page 178 is $^{126}/_{1,716}$. This calculation can be done by those who go to the effort of reading and understanding the optional section on theory, but it is sufficient to accept that the calculation can be done. The probability of the outcome given in the table on this page is $^{7}/_{1,716}$. Overall, the combined probability of an outcome at least as extreme as ours in favor of women preferring margarine is then $^{126}/_{1,716} + ^{7}/_{1,716}$ or $^{133}/_{1,716}$. This is then our p value. However, there is a further complication here. This is the p value for a one-tail test. It is the p value if H_0 was as it was before, but H_a was that women tended to be margarine preferrers. These hypotheses would be valid if we believed that a reasonable starting point was that there was no gender distinction in these preferences, but that if this was in fact incorrect it would have to be in favor of women preferring margarine. Such an H_a might be appropriate if we believed that the only possible connection between gender and butter preferences was that women are associated with nurturing and are therefore perhaps more health conscious and hence into healthier, lower-cholesterol food.

If, however, our H_a was simply that there could be gender differences in either direction, then our p value would have to take into account the probabilities of all the outcomes at least as extreme as the outcome obtained, including outcomes where instead of too few there are too many females (and hence too few males) in the butter-preferrer category compared to the numbers we expect from the fact that about half of all the people are butter preferrers. Calculation shows that the outcome with two male and five female butter preferrers out of six men and seven women occurs more commonly than the outcome we actually obtained of five male and two female butter preferrers. We therefore don't count the probability of the outcome with two male and five female butter preferrers: It is not as extreme a result as the one we actually obtained. However, the outcome with one male and six female butter preferrers is a less common outcome than the outcome we actually obtained, as is the even more extreme zero male and seven female butter preferrers. Calculation shows that these latter two outcomes have probabilities under H_0 of $^{42}/_{1,716}$ and $^1/_{1,716}$. The two-tail p value is then the sum of all the probabilities of all individual outcomes at least as way out and unlikely as our actual outcome of five male and two female butter preferrers out of six men and seven women. It is therefore $^{133}/_{1,716} + {}^{42}/_{1,716} + {}^1/_{1,716}$ or $^{176}/_{1,716}$. In practice, it is not necessary to calculate all these individual probabilities: A computer program does all the work automatically.

However, leaving everything to the computer can lead to mistakes. There can be problems to do with computer calculation of the p value for one-tail tests. You can be stung by the tail, so to speak. Consider the following example from real life where an experiment or "clinical trial" was carried out to assess the value of antibiotics in preventing infection after a certain surgical procedure. The results were that eight out of fifty-three people who received antibiotics got infections and three out of fifty-seven who did not receive antibiotics got infections. The computer analysis of this clinical trial gave a one-tail p value of 0.080 and a two-tail p value of 0.115. Since antibiotics are known to be effective against many infections, we might not want to strongly hold on to the idea H_0 that they are of no benefit. We therefore might set our benchmark p value here above the traditional 0.05. We also note that a one-tail test seems appropriate here. If antibiotics make a difference, it would be expected to be a positive difference. Therefore, at first sight the result 0.080 for the one-tail p value might make us think that we have reasonably convincing evidence that antibiotics are effective in this situation. However, in doing this calculation the computer automatically assumes that the appropriate p value is less than half. In particular, the computer is working out the combined probability under H_0 that with a total of eleven people getting infections it would be three or fewer of the nonantibiotic people who got infections. This is, of course, an answer to the wrong question. Our H_a is that if antibiotics do anything they will be of benefit in preventing infection. The computer is implicitly responding to the H_a that if freedom from antibiotics does anything it will be of benefit in preventing infection. Our question instead should be, "How often would chance alone lead us to getting as few infections as we actually did or even fewer infections in the antibiotic group compared to the nonantibiotic group?" Since we actually got more infections in the antibiotic group despite the fact that this group was a bit smaller than the nonantibiotic group, our answer to this question should be more than half the time. Put another way, we see that about 10 percent of the patients get an infection, but infection happens to about 15 percent in the antibiotic group and about 5 percent in the nonantibiotic group. Even if antibiotics were entirely useless, we should usually get results that look much better than these results. Calculation shows that 98.1 percent of the time, if antibiotics were entirely useless, we would get a result that looked as favorable or more favorable for antibiotics than our result. Our *p* value should therefore be 0.981, not 0.080. There is certainly no evidence in this result in favor of antibiotics, but if we entirely ignored common sense, did not bother to look at our figures, and simply looked at the *p* value, we might think that the experiment had come close to the "magical" 5 percent statistical significance value "proving" that antibiotics are worthwhile.

THE CHI-SQUARE TEST OF ASSOCIATION

The calculations involved in Fisher's exact test are difficult even for a computer when dealing with numbers greater than a few hundred. For this reason, and also because of the subtle philosophical objection discussed in the earlier optional section, an approximate test was developed. The test is called the chi-square test of association or simply the chi-square test (also written as the χ^2 test). The test has the advantage of applying not just to 2×2 contingency tables, but also to situations where there are more than two groups and/or categories. There is no widely used equivalent to Fisher's exact test in such situations. This use of the χ^2 test is dealt with later.

The detailed theory underlying the χ^2 test is complicated, but the actual calculations are simple. To explain the calculations, assume that one-third of all people surveyed are women and one-quarter of all people surveyed are butter preferrers. Our null hypothesis is that there is no gender effect on butter–margarine preferences. In that case, we would ideally expect that one-quarter of the women would be butter preferrers. Since one-third of all people in the sample are women, this would mean that a quarter of this third of the total or one-twelfth of the total should ideally be in the box "female and butter preferrer." This is just an ideal. Even if the null hypothesis is correct, chance alone will often mean that the number will not be exactly one-twelfth of the total. Indeed, if the total was not a number divisible by twelve we couldn't possibly achieve this ideal. Regardless of this, we call this ideal number the expected value and we denote it by *E*. The actual observed number we denote by *O*. We then calculate the quantity

$$\frac{(O-E)^2}{E}$$

We can repeat this calculation until we have dealt with all four entries in the 2 × 2 contingency table (male and butter preferrer, female and butter preferrer, male and margarine preferrer, female and margarine preferrer). We then add together all four quantities

$$\frac{(O-E)^2}{E}$$

There is theory that tells us that if H_0 is correct, the value we have obtained is approximately a value from a probability distribution known as a χ^2_1 distribution (pronounced "chi-squared distribution with 1 degree of freedom"). If the observed values are suspiciously far from the expected values (remembering that we expect O to be close to E under H_0), then $(O-E)^2$ will be large and we will get a suspiciously large number from a χ^2_1 distribution. More precisely, if the sum of the values

$$\frac{(O-E)^2}{E}$$

for all four boxes is x, then the p value is approximately the chance of obtaining the value x or larger from a χ^2_1 distribution. In other words, if H_0 were correct, $1-p$ of the time our Os would be a closer match to our Es than we have found in our case. If p is tiny, so that $1-p$ means "nearly always," we could say that if H_0 were true we would "nearly always" get our Os matching our Es better than we have. This would suggest that H_0 is not true, though, as always, we would have to judge the appropriate benchmark p value in light of common sense. It should also be noted that the theory on which the test is based is only approximate and the approximation is regarded as inadequate when the expected number in any box is less than 5. When numbers are large, the χ^2 test and Fisher's exact test give similar p values. The p values obtained from a χ^2 test are those for a two-tail test and need to be halved if a one-tail test is appropriate. Note again the caution givenon pages 180 and 181 and in the section on one-tail and two-tail tests on pages 75 and 76 regarding p values for one-tail tests when the data point in an unexpected direction.

For example, let us apply the χ^2 test to our gender butter–margarine preference example from page 178. Since $7/13$ of the sample are butter preferrers and $6/13$ are male, we ideally would expect $7/13$ of $6/13$ of the total of thirteen people to be male butter preferrers. This is about 3.23 people (more briefly, the formula for expected value is

$$\frac{\text{row total} \times \text{column total}}{\text{overall total}}$$

Similarly, we would expect 3.77 to be female, and among the margarine preferrers there would be 2.77 males and 3.23 females (note that once we have calculated one of the expected values, we can get all the other expected values using the fact that the expected values have the same group and category totals as the observed values). We then calculate

$$\frac{(3.23-5)^2}{3.23} + \frac{(3.77-2)^2}{3.77} + \frac{(2.77-1)^2}{2.77} + \frac{(3.23-5)^2}{3.23} = 3.90$$

The computer or a table of χ^2_1 values shows that numbers at least this large from a χ^2_1 distribution occur a little less often than 5 percent of the time ($p = 0.0484$). We could then say, "We have evidence that is statistically significant at the 5 percent level that there is a gender difference in butter–margarine preferences." We can compare this result with our two-tail p value of $^{176}/_{1,716}$ or 0.103 obtained from Fisher's exact test. Unfortunately, though, our χ^2 test calculations are not valid because some (in fact all) of our expected values are less than 5. The approximations underlying the χ^2 test make it unreliable with these small numbers. We should therefore use the p value obtained from Fisher's exact test.

ODDS RATIO AND RELATIVE RISK

Fisher's exact test and the chi-square test of association test the null hypothesis that there is no relationship between which group a person is in and which category that person belongs to. If we decide to reject this null hypothesis, it is desirable to have some measure of the strength of the association between being in a particular group and having a particular category. There are two such measures: the odds ratio and the relative risk. The p value is also an indicator of the strength of the association, but it is also affected by the numbers in our sample. For example, a p value may be 0.01 because we have a large sample and there is a weak association, or because we have a small sample and there is a strong association.

The relative risk measure comes from consideration of a contingency table of the accompanying form. A table with such headings could apply in two medical research situations.

	Disease	No Disease	Totals
Factor present	a	b	$a+b$
Factor absent	c	d	$c+d$
Totals	$a+c$	$b+d$	$a+b+c+d$

One situation is that of a prospective trial in which people who are all initially disease free are followed until a certain number contract the disease.

Some of the people have some possible risk factor (or protective factor) and others do not. The estimate of the risk of disease over this time in those who have the factor is then $a/a+b$ and in those who don't have the factor is $c/c+d$. A summary measure that will be of interest in judging the importance of the factor in the risk of the disease is a measure of how big the risk of disease is in those who have the factor in comparison to how big the risk of disease is in those who don't have the factor. This is the quantity

$$\frac{a/(a + b)}{c/(c + d)}$$

called the relative risk. The relative risk gives a good summary of the results of a prospective trial.

The other medical research situation giving rise to a contingency table with these headings is a retrospective trial. In the retrospective trial some diseased people are chosen and some nondiseased people or "controls" are chosen for comparison. All the people are then asked about their previous exposure to some possible risk factor or protective factor. While the table may look the same as a table giving the results from a prospective trial, there is a major difference. It is not possible to calculate risk in this situation. Risk in a prospective trial is the proportion of diseased people out of the total of diseased and nondiseased people, but in a retrospective trial the numbers in these two groups is entirely arbitrary. Even the ratio of the "risks" according to exposure to some factor, the relative risk, is not valid. If we choose few controls, then the diseased people may form a very large majority in the group exposed to the risk factor but may also form quite a large majority even in the group not exposed to the risk factor. The two "risks" will then both be close to 1, and so the relative risk measure will be close to 1. On the other hand, we could gain a different impression if we choose a large number of controls. For example, say we choose 20 controls, with the remaining figures as given in the accompanying table.

	Disease	No Disease	Totals
Factor present	100	10	110
Factor absent	50	10	60
Totals	150	20	170

The "relative risk" would be

$$\frac{100/110}{50/60} \approx 1.09$$

If instead we choose 2,000 controls and the numbers were otherwise in the same proportion, we would have the accompanying table and the "relative risk" would be

$$\frac{100/1110}{50/1050} \approx 1.91$$

	Disease	No Disease	Totals
Factor	100	1000	1100
No Factor	50	1000	1050
Totals	150	2000	2150

For a retrospective trial what we need in place of relative risk is a measure of the relationship between disease state and factor exposure that doesn't depend on the proportion of controls that we use. The measure that fits the bill is called the odds ratio.

To explain the odds ratio, we first must explain "odds." Odds are an old measure of probability still used in gambling, particularly gambling on horse races. The odds are 5 to 1 against a horse winning, or $5/1$, if it is five times more likely that the horse will lose than it will win. Some thought shows that this means that the chance of losing is $5/6$ and the chance of winning is $1/6$. In the case of the table on page 184, the odds of disease are $100/10$ for those with the factor and $50/10$ for those without the factor. The ratio of these two is the odds ratio. It is

$$\frac{100/10}{50/10} = 2.00$$

For the case of the table at the top of this page, the case of 2,000 controls, the odds ratio is

$$\frac{100/1000}{50/1000},$$

which again equals 2.00.

After simplification, the odds ratio for the 2 × 2 table on page 183 is

a	b
c	d

$$\frac{a \times d}{b \times c}.$$

An odds ratio of 1 means that there is no relationship between classification into rows and classification into columns: The odds of being in a particular column are not affected by which row a person is in.

The odds ratio has some additional advantages. We have previously noted that there is a symmetry in a 2×2 contingency table in that it may be arbitrary whether we define the groups or categories as the rows or as the columns. The odds ratio gives the same value for the relationship in a table regardless of whether we define either the groups or the categories to be the rows or the columns. The odds ratio makes sense as a measure of the relationship between the classification into rows and columns regardless of the context of the table. It applies to all 2×2 contingency tables, not just tables of the results of a medical trial on disease and risk-factor exposure. There is theory that allows us to apply confidence intervals to measures of the relationship between the classification into rows and columns in a 2×2 contingency table. This theory leads to fairly simple formulae when the relationship is expressed in terms of odds ratios.

Confidence Intervals on the Odds Ratio

Theory that is beyond the scope of this text shows that if the numbers in the boxes are sufficiently large, approximate confidence intervals can be worked out for the odds ratio. Recall that a confidence interval is the range of values of some underlying population parameter that could "easily" give rise to a sample with the observed data. By "easily" we mean that we would not reject a hypothesis that the underlying parameter was somewhere in the range. We define "easily" in terms of the p value that would make us reject the hypothesis. Here, a 95 percent confidence interval is the range of the underlying odds ratio that might exist in the population that could "easily give" the sample odds ratio. By "easily give" we mean that if we know the population odds ratio, our sample odds ratio is not unusual in the sense that odds ratios at least as far out as ours from the population odds ratio will occur by pure chance more than 5 percent of the time.

OPTIONAL

To give the formula for the confidence interval for an odds ratio, some further notation is necessary. Let z_q be a number such that the probability of a value chosen from a standard normal distribution being between $-z_q$ and $+z_q$ is q. For example, if q is 0.95, z_q is 1.96 because 95 percent of the probability in a standard normal distribution is between -1.96 and $+1.96$. The $100 \times q$–percent confidence interval for the odds ratio is then given by the following formula:

$$\exp\left[\ln(\text{Odds Ratio}) \pm z_q \sqrt{\frac{1}{a} + \frac{1}{b} + \frac{1}{c} + \frac{1}{d}}\right]$$

The symbols exp and ln stand for the mathematical functions of exponentiation and taking natural logarithms, functions that are standard on scientific calculators. For example, the 95 percent confidence interval is given by

$$\exp\left[\ln(\text{Odds Ratio}) \pm 1.96 \sqrt{\frac{1}{a} + \frac{1}{b} + \frac{1}{c} + \frac{1}{d}}\right]$$

In particular, say we were dealing with the table on page 184. The odds ratio, as we have seen, is 2.00. The formula tells us that the 95 percent confidence interval on this odds ratio is

$$\exp\left[\ln(\text{Odds Ratio}) \pm 1.96 \sqrt{\frac{1}{100} + \frac{1}{10} + \frac{1}{50} + \frac{1}{10}}\right] = (0.781, 5.12).$$

Previously our confidence intervals have been symmetrical about the estimate from the data. The nature of the calculations means that this is not the case for odds ratios, so 2.00 is not in the middle of the range 0.781 to 5.12.

END OPTIONAL

CLASSIFICATION IN MORE THAN TWO DIRECTIONS AND SIMPSON'S PARADOX

Imagine we tried out a new treatment for a serious disease in two hospitals, A and B, with the following results.

	died	lived
new	100	100
old	10	1

Results from hospital A

	died	lived
new	1	100
old	100	1000

Results from hospital B

At hospital A the odds of living with the new treatment are 1/1 and with the old treatment are 1/10, so that the odds ratio is 10 in favor of the new treatment. When the experiment was repeated in hospital B, the odds of living with the new treatment are 100/1 and with the old treatment are 10/1, so that again the odds ratio is 10 in favor of the new treatment. So far, so good, but then someone suggests combining the results of the two treatment trials to present the figures in an overall form, as shown here.

	died	lived
new	101	200
old	110	1001

Results from A and B combined

The odds of living with the new treatment are 200/101 and the odds of living with the old treatment are 1,001/110. This is an odds ratio of 0.22 in favor of the new treatment. In other words, it is an odds ratio of 1/0.22 or 4.6 against the new treatment. Which treatment would you rather have? Both sets of individual hospital figures are strongly in favor of the new treatment, yet the combined figures are strongly against. Which should be believed?

This is Simpson's paradox. The resolution of Simpson's paradox is to note that hospital B has a far better success rate overall. We will assume that they deal only with mild cases of the disease. Hospital B is also far more likely to be using the old treatment than hospital A. By combining the two tables, we would then be comparing the results of applying mainly the old treatment to mild cases of the disease with the results of applying mainly the new treatment to severe cases of the disease. The combined table is misleading. The new treatment is better.

Simpson's paradox can arise whenever along with classification into group (live or die) and category (new or old treatment) there is a third direction of classification (e.g., severe or mild disease) that is ignored in combining information. The third direction of classification is sometimes referred to as a *confounding variable*. A similar effect of confounding variables can occur in the situation of continuous measurements on two groups. In general, things become more complicated when there are more than two directions of classification or confounding variables. Dealing properly with such situations is beyond the scope of this text, but involves the topic of log–linear models and the Mantel–Haenzel test.

THE χ^2 TEST WITH MORE THAN TWO GROUPS OR CATEGORIES

The χ^2 test generalizes easily to the situation where there are more than two groups and/or categories. We might have m groups and n categories. We could then summarize our results in an $m \times n$ contingency table. The χ^2 test is applied in a similar way to the way it is applied in the case of a 2×2 contingency table. We use the same principle as in the 2×2 case to find the expected value in each box of the contingency table and then calculate the value of

$$\frac{(O - E)^2}{E}$$

for each box and sum all these values. Under H_0 the number we obtain comes from a χ^2_k distribution, and this gives us our p value as before. Here k is given by the formula (number of groups − 1) × (number of categories − 1) or (rows − 1) × (columns − 1). This formula relates to the number of entries we could alter independently of each other and still keep all the row totals and column totals fixed. This k is sometimes called the "degrees of freedom." As before, the test is approximate and should not be used if any E is less than 5. Where there are E values less than 5 it may be possible to avoid the problem by combining rows or combining columns.

For example, say we were interested in whether there was some relationship between women and men who form couples in terms of favorite primary colors. The results of this survey are given in the table.

Observed		women			
		red	yellow	blue	totals
m	red	25	10	5	40
e	yellow	9	11	10	30
n	blue	6	4	10	20
	totals	40	25	25	90

To test this, we might choose ninety women who have male partners and ask each of the women and each of their partners to specify which was their favorite color as displayed on a test sheet painted with the three primary colors. If four-ninths of the women chose red and one-third of the men chose yellow, we would expect one-third of the four-ninths of all women who chose red to have partners who chose yellow. This would mean that $4/27$ of the total (i.e., 13.33 couples) would be in the box "woman likes red and her man likes yellow" (the calculations are performed ignoring the impossibility of fractions, but we have rounded to two decimal places).

This is our expected value *E*:

Expected		women			
		red	yellow	blue	totals
m	red	17.78	11.11	11.11	40
e	yellow	13.33	8.33	8.33	30
n	blue	8.89	5.56	5.56	20
	totals	40	25	25	90

The survey gives us our *O* or observed value. We calculate

$$\frac{(O-E)^2}{E}$$

for all nine boxes similarly and add to give a number *x*. If there is no relation-ship between partners in terms of primary-color preference, this figure *x* would ideally be zero, for the *O*s would match all the *E*s, but simply as a result of random chance we will usually not have all the *O*s matching all the *E*s exactly.

Nevertheless, if there is no relationship between partners in terms of primary-color preference, the *O*s and the *E*s are not usually a bad match, so the sum of the

$$\frac{(O-E)^2}{E}$$

for the nine boxes is not usually a large figure. The theory says that if there is no relationship, the value obtained for *x* will be approximately a value from a χ^2_k distribution. Here *k* is 4, as there are three rows and three columns: $(3-1) \times (3-1) = 4$. Tables or a computer will show values as large as *x* or larger occur only *p* of the time. This is then our *p* value. It is telling us that if there was no relationship between color preferences of partners and it were chance alone leading to the mismatch between *O*s and *E*s, there is probability *p* that the mismatch would be at least as bad as the observed mismatch. If this *p* value is small, a better explanation for the mismatch between the *O*s and the *E*s might be that it is not just random chance that is operating but that there is a real relationship between the color preferences of partners. Straightforward calculation (tedious by hand, easy by computer) shows that the sum of the

$$\frac{(O - E)^2}{E}$$

for the nine boxes is 13.93. The computer or tables then tell us that 99.248 percent of the time values chosen from a χ^2_4 distribution are smaller than 13.93. In other words, our p value is 0.00752. Then, as always, we have a choice. Either we believe that there is no relationship between the color preferences of partners—it just looks that way (in the sense of the Os being a particularly bad match to the Es) because a less than 1 in 100 chance came off—or else there is an underlying reason for the Os not to match the Es. The underlying reason is that there is some connection between color preferences among partners.

As a second example using the same data, let us assume that our ninety couples are representative of all men and women. We can then use the same data to ask whether color preferences in women are different from color preferences in men. Here are the relevant figures:

	red	yellow	blue	
women	40	25	25	90
	(40)	(27.5)	(22.5)	
men	40	30	20	90
	(40)	(27.5)	(22.5)	
	80	55	45	180

The expected values have been calculated and are given in parentheses. This gives a value of 1.01 for the sum of the

$$\frac{(O - E)^2}{E}$$

for the six boxes. We say that we have a χ^2 value of 1.01. The degree of freedom k is $(3 - 1) \times (2 - 1) = 2$. A computer or tables shows us that 1.01 is not an unusual value from a χ^2_2 distribution ($p = 0.603$), so we can conclude that the figures here give no convincing evidence that men and women differ in their color preferences. The differences from the ideal or expected values can easily be explained by chance, and, indeed, 60 percent of the time chance alone would lead to bigger differences than those seen here.

As a third example, say we were relating the four blood groups (O, A, B, AB) to eye color and we had the eye colors blue, green, hazel, and brown.

Now say one-tenth of the population had green eyes and one-thirtieth of the population had blood group AB, and say also that we had 1,000 people in our survey. The expected number of green-eyed people with blood group AB is then $^1/_{10} \times ^1/_{30} \times 1,000 = 3.333$. This expected number is too small. The χ^2 test will not be valid with fewer than 5 for any expected value. We could then "collapse" our categories "green" and "hazel" eye color. Instead of four eye colors we would have only three: blue, brown, and hazel or green. If one-tenth of the population had hazel eyes we would have two-tenths in the combined category "green or hazel," so then we would expect $^2/_{10} \times ^1/_{30} \times 1,000 = 6.667$ to be in the category "green or hazel" eye color and blood group AB. Our calculations would then be valid, as the expected value in this box is more than 5 (assuming there were also more than 5 in all the other remaining boxes).

SIMILAR TESTS BASED ON THE χ^2 DISTRIBUTION

There are several other tests that use the chi-square distribution in a similar way to the chi-square test of association. However, these tests apply to situations that are rather different from the situation of multiple groups and categories that applies in the case of the chi-square test of association.

McNemar's Test

One such test is McNemar's test. This test was previously described as part of the description of the sign test in Chapter 4. McNemar's test applies when we have pairs and each member of the pair can fit into one of two categories. In the case of the straightforward sign test, the pairs are usually pairs of measurements made on the same individual before and after an intervention and the categories are "better" or "worse." In the case of McNemar's test, the pairs are usually separate but related individuals and the two categories are something other than "better" or "worse." The ordinary simple calculations of the sign test can be applied directly to give an exact p value. However, the same theoretical approach alluded to in the optional discussion of the theory of Fisher's exact test can be applied to give an approximate p value in the case of McNemar's test. This approximate p value is obtained via a calculation that gives a number from a χ^2_1 distribution.

For example, say we wanted to know whether an antismoking program is more effective against smoking in women rather than men or vice versa. Our H_0 is that men and women respond to the program equally. We will assume that there is a lot of variation in people's ability to give up smoking and that a lot of this variation depends on people's social situation. We will then use a strategy that will minimize the tendency of this variation in social situation to swamp the effect we are looking for, the possible differences in the effectiveness of the campaign on men and women. In particular, we will look at husband and wife pairs, for both partners share the same social situation. Say that

1,920 husband and wife pairs, all of whom smoked, were given the antismoking program. Assume that the result was that in 1,000 pairs both husband and wife continued to smoke, in 900 pairs neither continued to smoke, in 15 pairs the wife stopped but the husband didn't, and in 5 pairs the husband stopped but the wife didn't. This scenario can be presented in the form of a table:

		husband		
		smokes	no smokes	
w i	smokes	1000	5	1005
f e	no smokes	15	900	915
		1015	905	1920

According to the theory behind McNemar's test, if H_0 is true then the quantity

$$\frac{(15-5)^2}{15+5}$$

is approximately a number from a χ^2_1 distribution. This quantity is 5 and the computer or tables show that values at least this large in a χ^2_1 distribution occur 0.0253 of the time. The conventional benchmark p value of 0.05 seems to me to be reasonable here, so I would conclude that it is reasonable to believe that the antismoking program is more effective in women than in men. Note that the calculation ignores the 1,900 cases where there was no difference between the men and the women in a partnership. Also note that the exact p value can be obtained by reasoning that if men and women are equally likely to give up smoking, then in those cases where just one of the pair gives up it is equally likely to be the man or the woman. In other words, we are dealing with the binomial random variable based on twenty cases and a probability of 0.5 of a case going one way or another; equivalently, we can say that we are dealing with a two-tail sign test based on twenty cases. Since we have obtained the value of 5 going one way and 15 going the other, the computer or direct calculation (see the sign test in Chapter 4) shows that the p value is 0.04139. We see that the figure we obtained using the traditional approach to McNemar's test with the χ^2_1 distribution is a rough approximation to the exact probability obtained by the conventional sign test.

For interest, let us look at the result of a statistical test that uses the same data but ignores the pairing. The data tell us that there are 1,015 men who continued to smoke and 905 who gave up, and 1,005 women who continued to smoke and 915 who gave up. These data can be analyzed by an ordinary chi-square test of association with the help of the table on p. 194 (the numbers in parentheses are the calculated expected numbers).

	men	women	
smoke	1015	1005	2020
	(1010)	(1010)	
no smoke	905	915	1820
	(910)	(910)	
	1920	1920	3840

Note that the numbers in this table are numbers of individuals, whereas the numbers in the table for McNemar's test are numbers of couples. The expected value in both the boxes in the top row is 1,010, since half of the population is of each sex so ideally we would expect half the smokers to be of either sex. Similarly, the expected value in both the boxes in the bottom row is 910. Calculation then gives a χ^2_1 value of 0.1045 and hence a p value of 0.747. Unlike our previous result, this is certainly not convincing evidence of a sex difference in the effectiveness of the antismoking campaign. We see that ignoring the pairing obscures our previous finding. By focusing on the variability in smoking status within couples, McNemar's test tells us that there is greater effectiveness of the antismoking campaign in women than in men. If we ignore pairing, this finding is overwhelmed by the factors that tend to make people change smoking status as a couple.

Chi-Square Test of Goodness of Fit

Sometimes we have theoretical reasons for believing that the data should follow a certain pattern and we want a test that tells us if it is reasonable to hold on to this belief when we obtain actual data. As a simple example, we may for some reason believe that all three primary colors should be equally likely to be a favorite color. Using the data on color preference from our earlier example, we see that red, yellow, and blue are favorites for a total of 80, 55, and 45 people, respectively. The theory of equal favorites would give us that 60 people each out of the 180 total would select red, yellow, and blue. This is an ideal or expected set of results. We want to know whether our actual figures are so far out from this ideal that it is not reasonable to put the difference down to chance. There is theory that tells us that under the null hypothesis of equal favorites, the sum of

$$\frac{(O-E)^2}{E}$$

for the three boxes will be a figure from a χ^2_k distribution. Here k, the degree of freedom, is the number of boxes − 1.

The sum of $\dfrac{(O-E)^2}{E} = \dfrac{(80-60)^2}{60} + \dfrac{(55-60)^2}{60} + \dfrac{(45-60)^2}{60} = 10.83.$

The computer or tables of a χ^2_2 distribution show that if the null hypothesis is true we will obtain a smaller number than this about 99.56 percent of the time. In other words, our p value is about 0.0044. It seems more reasonable therefore to believe that the primary colors are not all equally attractive. This is an example of the chi-square goodness of fit test.

The chi-square goodness of fit test can be used in more complicated cases. For example, we may know that if data come from a certain normal distribution, then a certain proportion of the sample should be in various categories defined by a range of values. For example, if we thought we knew there was a normal distribution with certain parameters describing the heights of people, then we would believe that certain proportions of people should have heights in the ranges 130 to 150 cm, 150 to 170 cm, 170 to 190 cm, or 190 to 210 cm. These give the expected numbers in the various categories, and we can compare these with observed numbers using the chi-square goodness of fit test. If the test yields a small p value, there are two possible explanations. On the one hand, our assumed normal distribution may be correct, but as a result of a chance that occurs a proportion p of the time we obtained observations that were a poor match to the true distribution. On the other hand, it may be more reasonable to believe that the population from which we have made our observations does not have a distribution of heights that matches our assumed normal distribution. As for the chi-square test of association, the calculation is not valid if the expected numbers in any of the boxes is less than 5. If we do not know the parameters of the normal distribution and our hypothesis is simply that it is a normal distribution with parameters to be estimated from the data, the theory tells us to deduct a degree of freedom for every parameter estimated. With four height categories, two parameters to be estimated from the data to describe the normal distribution, and a further degree of freedom to be deducted (to allow for fixed totals), we would be dealing here with a χ^2_1 distribution.

SUMMARY

- Fisher's exact test gives a p value that tells us how difficult it would be for chance alone to lead to an apparent association between two methods of classifying individuals that is at least as strong as the association present in our sample.

- The odds ratio measures the strength of the association evident in the sample, and confidence intervals can be calculated for the population odds ratio.

- The χ^2 test involves approximations, but can be used in place of Fisher's exact test when expected values in every box are greater than 5. It has some theoretical and practical advantages in comparison to Fisher's exact test.

- Unlike Fisher's exact test, the χ^2 test generalizes to situations in which there are more than two groups and/or categories.

- A test based on the χ^2 distribution can be used to assess goodness of fit to some theoretical distribution.

In short, with two groups and two categories and small numbers, use Fisher's exact test to see if there is convincing evidence that being in a particular group affects which category an individual is assigned to. With more than two groups or more than two categories or if numbers are large, use the χ^2 test.

QUESTIONS

1. The careers of 50 males who at the age of eighteen years were over 180 cm tall were compared with those of 100 males who at the age of eighteen years were under 170 cm tall. It was found that 12 men in both groups became administrators. Do you believe that this study provides reasonably convincing evidence that height is a factor in determining a man's chance of becoming an administrator?

2. A number of farms are selected at random from a list of farms. Thirty are family farms, and of these, ten are classified as being severely eroded. Twenty farms are owned by absentee landlords, and of these, fifteen are classified as being severely eroded. Does this data provide reasonably convincing evidence for an association between absentee ownership and erosion?

3. A drug thought to increase strength was given to three men, with nine other men acting as controls. The results in arbitrary units were 19, 20, and 21 for the men with the drug and 10, 11, 12, 13, 14, 15, 16, 17, and 18 for the men without the drug. The drug was also tried on nine women, with three women acting as controls. The results were 5.5, 6.0, 6.5, 7.0, 7.5, 8.0, 8.5, 9.0, and 9.5 for women with the drug and 4.0, 4.5, and 5.0 for women without the drug. Perform appropriate tests to see if there is convincing evidence that the drug works for men and for women. Combine the results for both men and women and repeat your statistical test. Explain your result.

4. A survey was conducted in which people were asked their religious affiliations and also asked whether they regarded preservation of the natural environment as a top-priority issue. Of thirty-three protestants surveyed, ten rated the environment as a top priority; the same was true of nine out of nineteen catholics, four out of ten moslems, and twenty-eight out of forty-one atheists. Does this survey provide reasonably convincing evidence that the priority accorded to the environment is related to religious affiliation?

5. A researcher is interested in a possible association between having a carpeted bedroom and asthma. Accordingly, she identifies 200 families, each with two children where just one of the children has asthma. She records the information that in 130 families both children's bedrooms are carpeted and in 40 cases neither children's bedroom is carpeted. In the remaining cases only one child's bedroom is carpeted and in 8 of these it is the nonasthmatic child's bedroom that is carpeted. Perform the most appropriate test on these data. Do you believe there is an association between asthma and carpet in the bedroom? Perform a Fisher's exact test (or chi-square test of association) on these data. Why are the results different?

6. To test if a die is fair, it is thrown 600 times. It lands on the numbers one to six, 93, 98, 107, 102, 91, and 109 times, respectively. Perform an appropriate test and discuss whether you believe that the die is fair.

CHAPTER 9

Statistics on More Than Two Groups

Most of the work on statistical tests so far has concerned comparing measurements between two groups. The last chapter considered situations where there were more than two groups, but only dealt with categorical measurements, deciding which category each member of each group belonged to. In this chapter we consider situations in which there are more than two groups and continuous measurements are made on each of the members of the groups. In the first part of this chapter we consider situations in which there are a limited number of groups and there is no natural ordering of the groups. The groups might correspond to people with a disease who are grouped according to which of several different medications they have received, or plots in a field grouped according to the type of fertilizer they have been treated with, or people grouped according to their eye color. This is the topic of ANOVA.

In the second part of this chapter we consider a situation in which the number of groups can in a sense be infinite. An example where the number of groups are in a sense infinite occurs when we ask about the relationship between weight and height. To answer this question we could divide people up into groups according to whether their heights are in the range 130 to 150 cm, 150 to 170 cm, 170 to 190 cm, or 190 to 210 cm and then measure their weights. However, we would do better if we divided height more finely than into groups of 20 cm. The best approach is to go further and deal with all the heights that occur separately. Since the number of possible heights are unlimited, we are in a sense dealing with an infinite number of groups. As well as the number of potential groups being infinite, another feature of this problem is that the groups are not separate unrelated categories but have a natural order. This work will be dealt with later in the chapter under the heading, "Regression."

ANOVA: TESTING THE NULL HYPOTHESIS THAT THE MEANS OF MORE THAN TWO NORMALLY DISTRIBUTED GROUPS ARE THE SAME

Motivation

We may be interested in comparing the effect on the growth rate of plants of more than two different types of fertilizer. To be specific, imagine that we wanted to examine the effect of eight different types of fertilizer, which we will call A, B, . . . , H. We could deal with this problem by reducing it to a number of different comparisons between the effects of two different types of fertilizer. We know how to do this: Use the independent samples t test. We could then make every possible pairwise comparison. In other words. we would compare A with B, A with C, A with D, A with E, A with F, A with G, A with H, B with C, B with D, . . . , G with H. However, this would be tedious: $^8C_2 =$ 28 pairwise comparisons are possible.

There would be another problem with this approach. Traditionally, we decide if a difference is "statistically significant" by asking how easy would it be for coincidence alone to explain differences in the averages of two groups. If we have to rely on a long coincidence to explain differences, we decide that coincidence is not a good explanation, but instead there are real differences due to the different fertilizers and not just due to chance. Our examination of twenty-eight comparisons between pairs of fertilizer becomes in effect twenty-eight different searches for a long coincidence. In the presence of randomness, if we look long enough for a long coincidence we will find it. If here we take a long coincidence to mean the traditional benchmark p value of 0.05 or 1 in 20, we are giving ourselves twenty-eight attempts at finding a 1 in 20 coincidence. It seems likely that chance alone would lead us to find such a coincidence, and using the benchmark of $p = 0.05$ we would then unjustifiably declare the coincidence to be "statistically significant." One approach to dealing with this situation is to lengthen what we consider to be a sufficiently long coincidence for statistical significance when we make each comparison. In other words, we reduce the benchmark p value used in each comparison so that overall, by the time we have finished twenty-eight comparisons, if random chance alone is operating we have the required small probability (e.g., 0.05) of finding a statistically significant difference for one or more of the comparisons. Unfortunately, it is difficult to calculate how much the benchmark p value used for each individual comparison should be reduced. The calculation is made difficult by the fact that the various comparisons are not independent of each other. If it has been found that chance alone could very easily explain differences between group A and each of the other groups B to H, then it is more likely that each of the groups B to H are similar enough to each other for chance to very easily explain their differences. One approach is to use a very rough approximation that ignores such difficulties. In the situation here of twenty-eight comparisons this rough approximation tells us to require that each comparison be tested for statistical significance using a benchmark *p* value of 0.05/28. The name Bonferroni is attached to this approach, though Bonferroni showed that this approach would lead to an overall *p* value of less than 0.05.

Philosophy of ANOVA

The pairwise comparisons approach is awkward. A better approach is a method that deals with all the data at once. The method is called *ANalysis Of VAriance* or ANOVA. We start with the null hypothesis that all the data in the various groups come from the same normal distribution. Figuratively, we can think of numbers chosen at random from a normal distribution being written down on pieces of paper and then placed in a hat. There are other hats containing numbers drawn from different normal distributions. A hat is chosen and some pieces of paper are pulled from this hat and piled into a group. The process is repeated several times to create several groups. It is quite likely, but not certain, that in choosing a hat to make each group we keep on choosing the same hat. Our question is, "Do all the numbers in all the different groups come from the same hat?" Our null hypothesis is that they do. We look for evidence against this null hypothesis. To do this we start by assuming that the null hypothesis is true (all the numbers in each of the groups all come from the same hat). We then focus on measuring the variability of the individual numbers (How scattered are the numbers in the hat?).

We use two different methods of estimating this variability. The first method is to estimate the variability within each group and form a pooled estimate of the underlying variability of the individual values. The second method is to estimate individual variability from the amount of variability of the group means. If the null hypothesis is correct, the results of the second method divided by the first gives a value which should be about 1, since both methods are methods of measuring the same quantity: the individual variability. However, if the group means are scattered, not only as a reflection of random variability of the individual values but also because of real differences between the true means of the various groups (the numbers in the different groups come from different hats), then estimating the individual variability using group means will generally give an overestimate. The individual variability overestimated from the scatter of group means divided by the individual variability as estimated by the scatter of individual values within groups is then likely to be considerably larger than 1. This ratio of the two measures of individual variability is known as an *f* ratio. If we had thought that all the numbers came from the same hat, an *f* ratio of more than 1 tells us that the group means are more widely scattered than we would have expected given the amount of variability we see within the groups.

Theory that is too complicated to be described here tells us that if the null hypothesis is true, the *f* ratio is a value from a particular probability distribution known as an F distribution. If the f ratio we obtain is a value so far above 1 that the F distribution shows values this big or bigger "hardly ever" occur, then we reject the null hypothesis that the group means are all the same. What we actually obtain here is a probability of obtaining a value at least as large as our f if H_0 is true. This is our p value. If our p value is a probability that "hardly ever" occurs, it may be more reasonable to believe that the variability of the means overestimates the individual variability not because of a chance that "hardly ever" occurs but because there are real differences between the true means of the various groups. Once again, "hardly ever" is quantified by our benchmark p value.

There are a whole family of different F distributions appropriate here, depending on the number of values used in both of the estimates of variability (the degrees of freedom). The notation used is $F_{\nu_1 \nu_2}$. Here, ν_1 is the degrees of freedom (number of estimates of variability) used in the numerator of the ratio of the two estimates of individual variability and equals the number of groups minus 1, and ν_2 is the degrees of freedom (number of estimates of variability) used in the denominator and equals the sum of numbers minus 1 in each of the groups. In practice, the calculations are lengthy and usually performed on a computer, and the computer attends to details about these degrees of freedom.

ANOVA Assumptions

ANOVA is a good method for making a decision between two hypotheses. One hypothesis H_0 is that all the values in all the groups are chosen from the same normal distribution. The other hypothesis is that all the values in all the groups are chosen from normal distributions with different means but all sharing the same value for their standard deviation. These two options can be called the ANOVA assumptions. Often, though, we do not want to make a decision between these two rather restrictive options. We often just want to decide whether or not it is reasonable to believe that the means of all the groups are the same, regardless of assumptions about normality and equal variances. Applying ANOVA when the assumptions are not met can give misleading results, as the two examples in the following optional sections show.

OPTIONAL ▬▬▬▬▬▬▬▬▬▬▬▬▬▬▬▬▬▬▬▬▬▬▬▬▬▬▬▬▬▬▬▬▬▬

Violation of the Normality Assumption

If the numbers in each of the groups are not normally distributed, the ANOVA calculation of a p value will not be valid. For example, if the numbers in all the groups are drawn "from the same hat" but came from a distribution that only took the values 0 and 1 with equal probability, then the basic probability laws of Chapter 3 can be used to work out how often the between-group variance will be positive but the within-group variance will be zero. In particular, there are 2^9 or 512 ways of choosing a 0 or a 1 nine times in a row, with all these ways equally likely. The first three choices, the second three choices, and the third three choices can be taken to be the groups. Simple counting then shows that there is a 6 in 512 chance that the between-group variance will be positive but the within-group variance will be zero. The two estimates of individual variability, one based on variability among group averages and the other based on variability within each of the groups, will therefore be positive and zero, respectively. The f ratio is then a positive divided by zero; in effect, it is infinity. In other words, with this distribution the p value corresponding to what is in effect an infinite f ratio would be 6/512 or about 0.012, whereas the p value of an infinite f ratio if the numbers all came from a normal distribution would be zero. Similarly, it can be shown that if the numbers are drawn from the distribution that gives 0s and 1s with equal probability, the p value associated with an f ratio of 7.0 is 42/512 or about 0.082. If instead the numbers are drawn from a normal distribution, an f ratio of 7.0 would correspond to a p value of 0.027. In this example, if we incorrectly assume normality, we may decide more easily than we had intended to reject the null hypothesis when it is true.

Violation of the Equal Variance Assumption

In this example we show that if the different groups are normally distributed but are different in both means and variances, the ANOVA calculations may lead us to not choose the alternative hypothesis that the groups are different when it is clearly appropriate to believe that they are different. For example, say one group was widely scattered but with values centered on 60, and there were two other groups with very little scattering and one of these consisted of numbers very close to 40 and the other consisted of numbers very close to 80. If we round the numbers to two significant figures, a sample of three values from each of these three groups might then consist of the numbers 0, 60, 120 for the first group; 40, 40, 40 for the second group; and 80, 80, 80 for the third group. It is obvious to anybody looking at these figures that the three groups do not look like three groups of numbers drawn out of the same hat. Any sensible method of assessing group differences should find evidence that the groups are different. However, it turns out that in the ANOVA calculations both methods of estimating individual variation (with one method based on variation within groups and the other based on the variation of group averages) give exactly the same answer. In other words, the f ratio is 1.0 and the p value corresponding to this is 0.42, so according to ANOVA there is no evidence that the numbers come from different hats.[1]

Checks of the Assumptions

Since applying ANOVA when the assumptions are not met can give misleading results, it is desirable to check to see if there is evidence against the

ANOVA assumptions using box plots and preliminary statistical tests. Such tests are available, but are often omitted. In the two examples here it was the example in which there was violation of the equal variances assumption where the ANOVA analysis seemed to be most misleading. In this example we obtained a p value of 0.42 where common sense suggested that we should have obtained a p value indicating strong evidence against the null hypothesis. In the example where there was violation of the normality assumption, the ANOVA calculations underestimated the true p value, but the true p value and the p value calculated using ANOVA were both small. This suggests that it is the violation of the equal variances assumption that can lead to decisions about hypotheses that are more obviously inappropriate. In practice, checks to see if there is evidence against the equal variance assumptions are sometimes performed, but tests to see if there is evidence against the assumption of the normality of each group are rarely done.

The common test for equal variance is called the Levene test. This test uses the principle that equal variances within each group by definition means equal values for the average of the square of the differences between each value and the group average. This then implies equal values on average for the size—disregarding plus and minus signs—of the difference between a value and its group average. The Levene test then does a preliminary ANOVA to see if there is evidence against the assumption that the size of the difference between a value and its group average is on average the same in every group. If a Levene test shows that it is not reasonable to believe that the groups have the same variance, then we have two options:

1. The most commonly used option is to hang on desperately to the ideas of testing null hypotheses. With this option we admit that the variability in the different groups are different. We dismiss our null hypothesis that the group variances and averages are the same and instead produce another null hypothesis that states that the group averages are the same but we don't know anything about the group variances. We then look for convincing evidence against this new null hypothesis. We can no longer use the standard ANOVA on the data as it stands. Instead we can use data transformation as described at the end of Chapter 5 or we can use a nonparametric equivalent to ANOVA known as the Kruskal–Wallis test. This test does not assume equal variances. It also does not use the actual data values, but only uses their rank order. Since it does not use all the information we have available it is a less powerful test.

2. The alternative option is to argue that if we are convinced that the groups have different measures of variability it is hardly reasonable to believe that they all have exactly the same means. Almost any intervention that affects variability could be expected to have at least some effect on means. We can therefore finish at this stage and say that we believe that the numbers in the different groups come from different hats and so can be presumed to have different averages. Although this option accords more with common sense, it is not the one commonly used.

END OPTIONAL

ANOVA without Checks of the Assumptions

Often checks of the assumptions underlying ANOVA are omitted. ANOVA is then applied in ignorance of whether the required assumptions are met. While this approach isn't ideal, it is commonly used in practice (see the fertilizer example later in this chapter for a typical application of ANOVA). Perhaps the difficulties in the philosophy of hypothesis testing justify not adhering closely to ideals. Alternatively, the equivalent nonparametric test can be used. As mentioned in the optional sectional earlier, the equivalent nonparametric test is known as the Kruskal–Wallis test and does not require the assumptions required by ANOVA.

Contrasts and Post-Hoc Tests

If the end result of an ANOVA is that there is evidence that there are real differences between the groups, there are further questions that we can ask. For instance, in the case of the eight different fertilizers, we may want to know if just one of the fertilizers gave an outstanding result and if it would be reasonable to believe that the other seven fertilizers all gave the same growth rate. There is a philosophical problem here. If we ask this question before doing the experiment and we have a particular fertilizer in mind, the correct approach is an independent samples *t* test, testing this fertilizer against the average of all the others. In the language used in ANOVA such testing is called a "contrast." If we ask this question after seeing the results and without having a particular fertilizer in mind, we need to remember that with eight groups, one is bound to be the biggest and the biggest of eight could easily be outstanding mainly as a result of chance factors. ANOVA may have told us that the groups are too different in their averages for these differences to be put down to variability of individual plants. This does not necessarily mean that the biggest is the one that is different. For example, it could be that the eight different fertilizers are in two groups of four, a group of inferior fertilizers that all have identical small effects on plant growth rates and another group of superior fertilizers that all have identical big effects on plant growth rates. The fertilizer with the biggest effect may stand out simply because, by chance alone, the biggest of four is likely to be unusually big. To try and tease out answers to questions like this is the subject of post-hoc tests (in other words, tests after the event of deciding that there are real differences somewhere between the fertilizers).

There is a variety of post-hoc tests with different philosophical approaches. The simplest approach is multiple *t* tests of all possible comparisons with the use of Bonferroni adjustment to the benchmark *p* value to choose which pairwise comparisons are "significantly" different. However, in many situations the use of post-hoc tests would come close to ignoring common sense. Our null hypothesis that eight different fertilizers all had the same effect on growth rate was already close to the absurd. If fertilizers are different then they should have different effects on plant growth rates. If we have found that chance alone cannot reasonably explain the differences in the performance of the eight different fertilizers, why should we have fallback positions? By fallback positions we mean retreating to the belief that, say, seven of them are the same and just one is outstanding, or that they are in two groups of four and performance within each group is identical. We might be able to have one or other or both of these fallback positions because post-hoc tests show that chance could then easily explain the remaining differences. If we are forced to drop a null hypothesis that was barely reasonable, that the different fertilizers all had identical effects, it seems more reasonable to now believe that different fertilizers all have different effects rather than believe in groupings simply because chance could then easily explain remaining differences.

ANOVA Example

ANOVA calculations are generally lengthy and most conveniently performed by computer. The computer was given data on the growth rate of plants treated with different fertilizers. The growth rates are expressed in arbitrary units and are given in the accompanying table.

Fertilizer A	Fertilizer B	Fertilizer C	Fertilizer D	Fertilizer E	Fertilizer F	Fertilizer G	Fertilizer H
71.5, 72.3,	70.2, 62.7,	66.4, 70.8,	83.8, 75.9,	80.1, 78.7,	76.1, 71.6,	81.2, 72.1,	81.5, 73.1,
68.2, 81.1,	63.2, 70.4	69.6	75.4	77.9	83.6, 74.7,	71.4	78.0
69.7, 71.8,					75.8		
65.1, 66.4							

The pds computer program written to accompany this book gives the one-line result $p = 0.003466$. In other words, if all the numbers for all the fertilizers were "pulled out of the same hat," then 99.6534 percent of the time the averages of the numbers for each fertilizer would be less widely scattered than the averages here. This can be predicted from the observed amount of scattering of the numbers for each fertilizer. The prediction is valid if all the numbers are drawn from the same normal distribution. It seems more reasonable to believe that the type of fertilizer affects growth rate rather than to believe that we got a suggestive result by sheer coincidence, the sort of coincidence which occurs only 0.003466 of the time. The "more information" button in the computer program gives means and confidence intervals for the means for each type of fertilizer. It also gives some intermediate results from the calculation and gives the result of the Levene test, which indicates that chance could "easily" account for observed differences in standard deviations in each group ($p = 0.66$). A separate Kruskal–Wallis test gives a p value of 0.00778. This tells us that even if we ignore the actual numbers and only look at the ordering of the thirty-two numbers from the eight groups, the results are still strongly suggestive of differences between the groups. The results are suggestive in that 99.222 percent of the time the ordered values would have values from the various groups more interspersed than they are here. The amount of interspersion is judged by a single figure calculated from each possible ordering, much like the Mann–Whitney U value is calculated in the case of just two groups.

Once again, we should note that the null hypothesis is close to absurd. Of course different fertilizers are going to have different effects on growth rates. The null hypothesis would only make sense if the different fertilizers were intended to have the same concentration of the same ingredients but were made in different factories whose production standards may or may not differ.

REGRESSION

Motivation

Often data consist of pairs of figures measured on each individual (e.g., weight and height), and we want to see if there is a relationship between the figures in each of the pairs. For the time being we assume that we are looking for a linear or straight-line relationship. In other words, we are assuming the following:

- If we are dealing with people who have average height, then on average they have average weight.
- If we are dealing with people who are 1 cm taller than average height, then on average they are β kg heavier than average weight where β is some constant.
- If we are dealing with people who are 2 cm taller than average height, then on average they are $2 \times \beta$ kg heavier than average weight.
- If we are dealing with people who are 3 cm taller than average height, then on average they are $3 \times \beta$ kg heavier than average weight.
- And so on.

In other words, we assume that for each extra cm in height, average weight goes up β kg, where β is some unspecified constant. This is the same as saying that plotting average weight against height on a graph gives a straight-line relationship. If average height is, say, 175 cm and average weight is 65 kg, a height of h cm is $h - 175$ above average and a weight of w kg is $w - 65$ above average. The relationship described here can then be written mathematically as $w - 65 = \beta \times (h - 175)$. This equation can be rearranged as $w = 65 - 175 \times \beta + \beta \times h$, which can be further simplified by writing the one symbol α in place of the constant $65 - 175 \times \beta$. The equation is then $w = \alpha + \beta \times h$. This is the normal way of writing an equation for the graph of a straight line where w

is the measure on the vertical axis and h is the measure on the horizontal axis. Traditionally, though, we use the symbols y and x for the dependent and independent variable, respectively, so traditionally the equation for a straight line is written $y = \alpha + \beta x$.

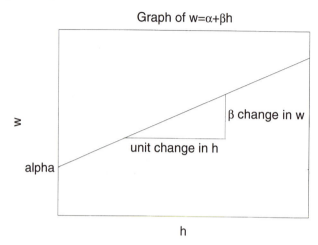

Graph of w=α+βh

Independent and Dependent Variables

Often we control one of the variables in that we make deliberate choices about the values that we are interested in. For example, if we are interested in the relationship between peoples' height and weight, rather than just choosing people at random and then measuring both height and weight we may choose people of particular heights and then measure their weights. We then think of height as the independent variable and weight as the dependent variable. Often, as here, the choice of independent variable is arbitrary: We could have just as easily chosen people by weight and then measured height.

Sometimes, though, one of the variables is something obvious, such as weight, and we choose people so that we have people with a range of different weights. We then make measurements of something that isn't externally obvious on the people of different weights. For example, we might measure their cholesterol levels. In this situation, weight is the independent variable; it depends on nothing but our choice. Cholesterol level is the dependent variable. It may depend on the weight of the people who were chosen.

We are dealing with situations in which there is variability and we assume that as well as a straight-line relationship there is additional variability or uncertainty. We incorporate uncertainty into our model by adding an "error term" e that takes different values for different individuals. If we use the term x_i for the measure of the independent variable on the ith individual, the term y_i for the measure of the dependent variable on the ith individual, and e_i for the error term here, we have the equation $y_i = \alpha + \beta x_i + e_i$.

Figure 9.1

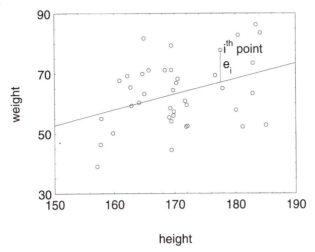

height

In this equation, x_i might be the height of the *i*th person we looked at, y_i might be their weight, and e_i is the error term appropriate for the *i*th person. In the theory that follows, we assume that the value taken by e_i is a value chosen from a normal distribution centered on zero. The error term e_i gives the vertical distance of the point on the graph from the "ideal" straight line. Figure 9.1 illustrates these ideas. Finding the ideal straight line is a process known as *regression* for reasons that will be explained later.

Why Use a Straight Line to Model the Relationship between the Variables?

Note that we don't know that the straight-line relationship is the correct relationship between height and weight. However, a straight-line relationship is a simple and plausible connection between the two variables in that an increase in one variable results in a proportionate increase in the other.

Even if the relationship is more complicated, so that the best fit to the data is given by a curve rather than a straight line, if we look at just a small part of this curve and magnify it it will look very similar to a straight line (see Figure 9.2).

A straight-line relationship may then be a reasonable description of the data over a limited range. Conversely, if we apply the formula for the fitted straight line to extreme values, it may well not be valid, since the true relationship between the two variables may only approximate a straight-line relationship over a limited range. Application of a formula that fits data over a limited range to values beyond that range is known as *extrapolation*. Extrapolation is unlikely to be valid, because the true average relationship between the two values, say height and weight, is unlikely to be exactly a straight line. Over a limited range the deviation from the straight line may be imperceptible, but it

Figure 9.2
Section of Curve under Extreme Magnification

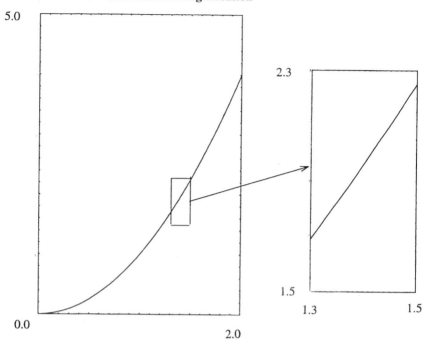

may become appreciable for extreme values. Unless we have data showing that the relationship is valid for extreme values, we should not assume that extrapolation is reliable. For example, in the weight versus height graph (Figure 9.1) the equation of the fitted straight line is $w = -25.48 + 0.52 \times h$, with w in kilogram and h in centimeters. Applying this formula to human children less than 49 cm in height would give the prediction that on average they have negative weight, which is of course impossible.

Defining the Straight Line of Best Fit and Assessing Its Accuracy

If we believe that the connection between the two measurements on individuals can be reasonably described by the equation $y_i = \alpha + \beta x_i + e_i$, we can ask some further questions. For example, how do we draw the straight line that best fits the data? Your first thought may be to simply draw, by eye and ruler, the straight line that looks like it gives the best fit. Compare this to the situation of a single measurement on each of a large number of individuals. In such a case we could mark the value for each individual along a line (a line graph; see Chapter 2) and then by eye put a mark where the values are centered. But we know how to do better than that with single measurements on individuals. Rather than just using our eye, we use calculations to find the mean (or some other measure of central tendency). We can then go further and obtain a measure of the reliability of the estimate of the mean. This is done by taking into account the number of data values and the variability of the data about the calculated mean. The end result is a confidence interval for the mean. Similarly, if we are given pairs of measurements on a sample of individuals we can obtain estimates for α and β by calculation, rather than by drawing, and so we define the line of best fit. Having obtained the estimates of α and β, we can obtain a measure of how reliable the estimates are in the form of confidence intervals. It can be shown that the line of best fit goes through the point representing the average of the x values and the average of the y values. The calculation of the slope of the line is rather lengthy and is usually done by computer.

The complete rules for calculating α and β will not be described here. Instead, the focus here is on interpreting the results of the computer calculations. A question that is of particular interest is, "Would it be reasonable to believe that β is zero?" If β is zero, our equation $y = \alpha + \beta x$ for the straight-line fit becomes $y = \alpha$. In other words, the value of x has no effect on the value of y. In the case of height and weight, this would mean that height has no effect on weight. In general, if $\beta = 0$, changes in the independent variable have no effect on the dependent variable.

Further Analogy with the Case of the Single Variable

The situation here is similar to our first example of continuous variables dealing with the height of Madagascans. We go back to the height of the Madagascans now and fill in some ideas that we glossed over. We then draw analogies with the calculations used in regression.

Our ideas on the height of Madagascans could be expressed by stating that the height of the ith Madagascan is $\mu + e_i$, where we assume that e_i is some error term obtained randomly from a normal distribution centered about zero. For valid calculation of an estimate of μ and the confidence interval for μ, we also need to assume the value e_i for every individual is drawn from the same normal distribution and that the value of e_i for any one individual has no effect on the value of e_i for any other individual. (If people of similar heights tended to mix with each other, then if the ith measurement was on an unusually tall person [$\Rightarrow e_i > 0$] this person's friend would also tend to be unusually tall and if the friend was the $i + 1th$ person sampled we would have a tendency for e_{i+1} to be greater than 0. This sort of problem would invalidate our calculations.) In brief, we state that our calculations will be valid if the e_i values are normal and identically and independently distributed. We don't know in practice if the e_i values meet these requirements. However, both experience and the central limit theorem suggest that often the e_i values won't be too far from normal, and unless we do something silly like take all of our sample from the one location that might happen to be the doorstep of the Madagascan dwarves association, the e_i values are likely to be identically and independently distributed.

If the heights of Madagascans are accurately described by the model $\mu + e_i$ with the e_i values normal and identically and independently distributed about zero, we first estimate μ by the mean of the sample and then we can go further and obtain a confidence interval for our estimate of μ. Finding a confidence interval is in effect an indirect answer to the question, "In light of the variation we see in individual values about our estimate of the mean, how reliable is this estimate of the mean?" The confidence interval doesn't answer this question directly, but gives the range of values that could be taken by the true value of μ that, given the individual variability, could still "easily" lead to our observed value of the mean. When there are n data values the calculation of the confidence interval involves the t_{n-1} distribution. The subscript $n-1$ relates to the fact that with two data values there is just one measure of variability and with n data values there are in effect $n-1$ measures of variability.

The situation with the calculation of α and β in regression is analogous to the calculations of μ for the Madagascans. The calculations give the best estimates of α and β from the data and valid confidence intervals, provided that the model $y_i = \alpha + \beta x_i + e_i$ is correct and the e_i values are normal and identically and independently distributed about zero. Confidence intervals can be found for α and β based on the same philosophy as finding confidence intervals for μ. We assess individual variability by measuring how much the data are scattered about the fitted line and use this to give an indirect answer to the question of the reliability of our estimates of the parameters of the line, the α and β. When there are n data values, the calculation of the confidence interval involves the t_{n-2} distribution. The subscript $n-2$ relates to the fact that with two data values there is no measure of variability about a straight line: A straight line can be drawn to fit exactly to any two points. With three data values there is in effect one measure of variability about the fitted line, and with n data values there are in effect $n-2$ measures of variability.

Asking if μ for Madagascans could be the same as the average American height involves hypothesis testing and the use of the t_{n-1} distribution. In the same way, the question "Is $\beta = 0$?" is answered by hypothesis testing and the use of the t_{n-2} distribution (recall that $\beta = 0$ implies the dependent variable is not affected by the values of the independent variable, as can be seen from the equation $y_i = \alpha + \beta x_i + e_i$).

As previously, the answers we can get are not quite the answers to the questions we really want to ask. The hypothesis test on the question, "Is $\beta = 0$?" gives a p value. This p value actually answers the question, "If the true value of β in the population were 0, how often would it happen that chance alone would lead to an estimated value of β from a sample at least as big as the β

found here?" We judge this chance by the variability in the data about the best straight-line fit and by the number of data points. As always, we need to use common sense, since we are not directly answering the primary question, "Is $\beta = 0$?" but instead obtaining an answer to a secondary, indirectly related question in the form of a p value.

If in our height and weight example we had a small sample, or a sample that just happened to contain a few very skinny giants and a few very fat dwarves, or a sample in which there was an unusual amount of variation in weight even after allowing for height differences, our test of the hypothesis $\beta = 0$ could give us a large p value, even as large as 1. The hypothesis test would be telling us that if β were 0, chance alone could easily explain any tendency seen in the sample for weight to increase with height. Of course, this does not mean that we should now believe that on average weight does not increase with height. Just because chance could "easily" explain any tendency for weight to increase with height does not mean that we should believe that chance is the explanation. To do so would defy common sense and the work of all those who have produced tables or graphs of average weights for people of various heights.

On the other hand, if H_0 is almost certainly true and H_a is almost certainly false, we should require an extraordinarily low p value before we start to believe that there is an association between the two variables. One study gave a p value of 0.000001 when a regression line was fitted to data over many years where one of the variables was the number of graduating Anglican ministers of religion and the other variable was the level of imports of Jamaican rum in the same year. Of course, common sense suggests that the explanation is likely to be that this is an instance where a 0.000001 chance actually occurred and if we were to continue observing these variables in future years we would find no relationship. My benchmark p value for believing that there was anything other than a chance association in the years observed would be smaller than 0.000001. However, this example also illustrates a further important point. If there was a real association it would almost certainly not be a result of Anglican ministers importing large amounts of Jamaican rum for their graduation parties, and it would almost certainly not be that the consumption of Jamaican rum induced Anglicans to take up a religious vocation, but it could conceivably be that the socioeconomic circumstances that somehow affect the number of graduating Anglican ministers also somehow affect the importation of Jamaican rum. Perhaps this possible connection is not so far-fetched. General economic circumstances could well affect both religious participation and importation of alcohol. In view of this, perhaps my suggestion about the benchmark p value is too extreme and we should allow the data to convince us of an association somewhat more easily. The important point here is that if it is decided that an association between two variables is too convincing to be put down to chance, then as explained in Chapter 7 on causality, the association may be because $A \Rightarrow B$, $B \Rightarrow A$, or $C \Rightarrow$ both A and B.

Complications That Do Not Have
an Analogy to the Single Variable Case

Violation of Assumptions

All the calculations in regression are only valid if the model $y_i = \alpha + \beta x_i + e_i$ is correct and the e_i values are normal and identically and independently distributed about zero. Since this model dealing with two variables measured on each individual is more complicated than the model $\mu + e_i$ for a single measurement on each individual, there is more chance that the model is not reasonably accurate and so there is more chance that the calculation of the best estimates of α and β and their confidence intervals are not accurate. For a start, we can have all the violations of the assumptions that can occur with the model $\mu + e_i$. In addition, the relationship between the average value of y and x, which is assumed to be a straight line, may in fact be a curve that differs appreciably from a straight line. The e_i values may also not be identically distributed. In fact, it is common for the size of e_i to depend on the size of x_i. In terms of kilograms, there is more variation in the weight of adult humans than in the weight of baby humans. Newborn humans vary from the tiniest premature babies that weigh less than 1 kg to exceedingly plump babies up to 5 kg or so. The mean, give or take 2 kg, covers the vast majority of newborns. On the other hand, only a relatively small minority of adults are within just 2 kg of average weight. This type of situation, in which the amount of variability in the dependent variable depends on the value of the independent variable, is said to be *heteroscedastic*.

If we are not dealing with a straight-line relationship or we are dealing with a heteroscedastic situation, then our calculations will not be valid. Inspection of graphs of our data and further analysis of the estimated e_i values obtained after fitting our straight line can indicate that the assumptions on which our calculations are based are unreasonable. We will consider only a few pitfalls that can be revealed by inspection of the graph.

Consider the four graphs in Figure 9.3. The data in all four graphs consist of the same number of points and give rise to the same regression line with the same confidence intervals for the parameters of the line. The top left graph portrays the ideal situation, with points scattered at random about a sloping line. The top right graph displays data points that fit perfectly to a parabolic arc. The bottom left graph displays data points, where all the points but one fit perfectly to a straight line that is not the regression line. The bottom right graph displays data in which all points but one have identical x values. Clearly, regression should only be used for the first graph. It is only for this graph that it is reasonable to believe that there is a general tendency for a straight-line fit but that individual points have random amounts of deviation about the straight-line fit. The assumptions underlying the calculation of the best straight-line fit do not seem reasonable in the case of the other three graphs. Instead, it is more

Figure 9.3
Graphs Illustrating Violations of Regression Assumptions

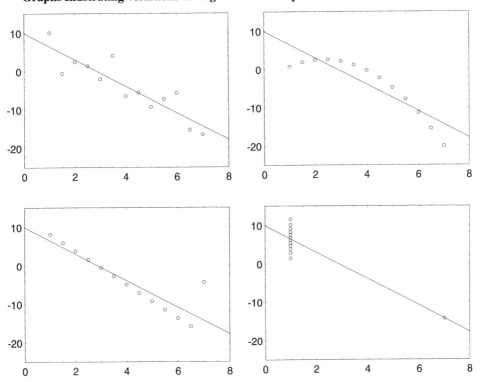

reasonable to believe that for the second graph we are dealing with an exact parabolic relationship, not an approximate relationship to a straight line. For the third graph it is more reasonable to believe that the anomalous point is an error, since the data otherwise fit exactly to a straight line. In the case of the fourth graph there is not enough information to give any check for the assumption that the variability of a point does not depend on its x value.

Procedure if Assumptions Are Violated

If examination of the graph of the data with the fitted straight line indicates that the assumptions are not valid but inspection does not suggest an obvious alternative such as a parabolic fit, we have two choices. First, we can use a nonparametric test for assessing the strength of the association between the values x_i and y_i. The most commonly used nonparametric test is known as Spearman's rho. It will be described later. The other option is to make arbitrary transformations of the x_i or y_i or both and repeat the regression calculations and graphing until it appears that with some transformation there is no evidence that the assumptions are violated. In other words, it may be possible to distort the *x* or *y* values or both in some way (e.g., by taking logs, squares, square roots, or whatever) so that a nonlinear relationship will "look like" a straight-line relationship and the estimated errors "look like" they are correctly behaved. If so, a straight-line fit to the transformed values of *x* and *y* can be calculated. This approach may seem rather contrived and one may be dubious about its validity. Nevertheless, it is quite commonly used.

Dependence of Estimates and Confidence Intervals for α *and* β

One problem with confidence intervals in regression is that the confidence intervals for α and β are not independent of each other. The value of α tells us how far up the vertical axis to start drawing the line (it is the value of *y* when *x* is 0). The value of β tells us the slope of the line. If in the weight versus height graph we thought that a value of α in the upper part of the confidence interval was more appropriate, the data would then suggest that the upward slope, the value of β, ought to be in the lower part of the confidence interval; if the line started higher up it would have to have a reduced slope to be a reasonable fit to the data. In particular, say we had confidence intervals for the estimate α = –25.48 and β = 0.52 in the equation $w = -25.48 + 0.52 \times h$. Let us also say we looked at this equation and decided that it didn't make sense with α = –25.48: A zero-height person ought to have zero weight, not negative weight. Particularly if 0 was in some reasonable confidence interval for α we might want to rewrite the equation as $w = 0.52 \times h$. Because the estimates and confidence intervals for α and β are not independent of each other, we would have to recalculate β and its confidence interval on the condition that α = 0. It turns out that the best fitting line of the form $w = \beta \times h$ is $w = 0.37 \, h$. This is illustrated in Figure 9.4.

Figure 9.4

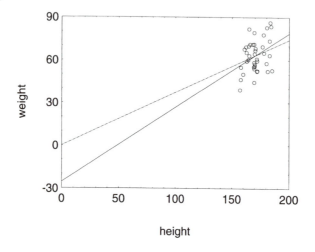

height

Figure 9.4 displays the same weight versus height data shown in Figure 9.1. Also displayed as a solid line is the previously calculated regression line

$$w = -25.48 + 0.521 \times h.$$

However, the axes have been rescaled to show zero. The dashed line is a new regression line calculated on the basis that a person of zero height has zero weight. It can be seen that moving the start of the regression line up the vertical axis by 25.48 units necessitates a decrease in the slope in order for the new line to give a reasonable fit to the data. This illustrates the fact that the estimates of α and β in the equation of the regression line $w = a + b \times h$, are not independent. Reassessment of one requires reassessment of the other.

Asymmetry between the Role of Independent and Dependent Variables and the Regression Effect

One problem with regression arises because the two variables are not treated equally. One of the two variables is taken to be the independent variable and the distance of the dependent variable from the fitted straight line is measured in terms of the vertical distance to the straight line. This definition is chosen because it is appropriate if we want to find the average value of the dependent variable that relates to one particular measure of the independent variable. We are then looking at all possible values of the dependent variable that have that particular value for the independent variable: The x value is fixed and we are considering all possible y values for that x value; that is, we are moving vertically, varying the y with the x fixed. However, this use of the vertical distance (as opposed, for example, to the use of a horizontal distance) introduces an asymmetry. This asymmetry can cause a number of problems.

For example, the regression formula fitted to the data displayed in Figures 9.1 and 9.4 has the formula for the relationship between average weight and height as $w = -25.48 + 0.52 \times h$. This formula tells us that the average weight of those exceedingly tall and rare individuals who have a height of 210 cm is about 84 kg. However, the converse is not true. The formula that relates average weight to a particular height cannot simply be put into reverse to give the average height for a particular weight. The average height of people who have a weight of 84 kg is not 210 cm: Overly wide normal-height people are much more common than properly proportioned exceedingly tall people. The average height of 84 kg people will be somewhat taller than average, but will be nowhere near 210 cm. These ideas lead to the regression effect.

Figure 9.5 may assist in understanding the problem. The concentric ellipses represent the density of points in a huge sample of people who have heights (h) and weights (w) measured. The more central the ellipse, the greater the density of sample points within it. The line marked "regression of w on h" is the line giving average weights of people of various heights. The vertical dashed line that almost touches the second-most outer contour shows that for the particular height 196 cm the numbers whose weights differ from the average for

Figure 9.5
The Regression Effect

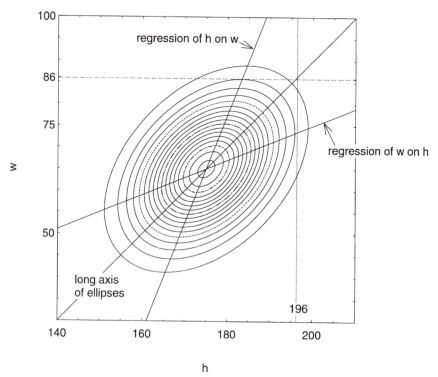

196 cm people fall off equally as we move away from the regression line in
either direction. This is shown by the fact that it is the same distance to the
outermost contour whether we go up in weight or down in weight. The line
marked "regression of h on w" is the line giving average heights of people of
various weights. The horizontal dashed line that just touches the second-most
outer contour shows that for the particular weight 86 kg the numbers whose
heights differ from the average for 86 kg people fall off equally as we move
away from the regression line in either direction. This is shown by the fact that
it is the same distance to the outermost contour whether we increase (go right)
in height or decrease (go left) in height.

The regression effect is a consequence of the asymmetry between dependent and independent variables in regression calculations. For example, it was
first noted in the nineteenth century that the sons of very tall fathers were on
average only moderately tall. This phenomenon was referred to as "regression
towards mediocrity," hence the name "regression" for this general area of statistics. A visual explanation can be obtained from Figure 9.3 by mentally rela-
belling and rescaling the axes. The horizontal axis could be labelled "height of

the fathers" and the vertical axis could be labelled "height of the sons." Both axes would have the same scale. The graph would then show that the average son of a very tall father tended to be smaller than the father, though there would be just as many very tall sons as there were very tall fathers. The verbal explanation is as follows: Assume that both genes and environment or genes and "luck" together determine height. Very tall fathers are mostly the result of "ordinarily tall" genes combined with luck ("very tall" genes are very rare, much rarer than the combination of ordinarily tall genes and luck). The very tall fathers pass on their mostly ordinarily tall genes to their sons. On average, the sons have average luck so end up merely tall rather than very tall. Put more simply, the association between the heights of father and son is not perfect: The father's height is only part of the explanation of the son's height; the other part of the explanation is chance.

OPTIONAL

As another example, say a casino puts on a new game. For a fee of just over $50, gamblers get to toss 100 coins. They get a return in dollars equal to the number of "heads" they throw. A number of gamblers who have tried the game once are concerned because they received an unusually low number of heads and hence dollars. Somehow a rumor starts that drinking orange juice will improve their luck. Amazingly enough, after drinking orange juice they all find that on trying the game again their scores have improved. Their results are listed here:

Gambler	a	b	c	d	e	f	g	h	j	k
Heads before orange juice	39	37	35	39	38	39	36	34	39	37
Heads after orange juice	54	43	49	52	51	48	50	56	45	51

Soon afterward, a number of gamblers who did very well on their first round of the game come back to the gaming table. They are very concerned to learn that the casino regards them as too lucky with heads to play the game in its current form. Instead, they will be given in dollars the number of "tails" that they throw. Again, somehow a rumor starts that drinking tomato juice will improve their luck by decreasing their heads score, conversely increasing their tails score. Amazingly enough, after tomato juice they all find that on trying the game again their heads scores have decreased as desired. Their results are listed here:

Gambler	l	m	n	p	q	r	s	t	u	v
Heads before tomato juice	67	61	62	65	61	63	61	62	64	57
Heads after tomato juice	49	47	53	52	55	46	48	54	49	50

A statistics student who witnesses all this is astounded by these results. He performs a sign test and a Wilcoxon signed rank test on both tables. All four tests give a *p* value of 0.00195. Furthermore, a *t* test on the first table gives a *p* value of 0.0000143 and a *t* test on the second table gives an even more convincing result of 0.00000898. The statistics student believes that, despite his prior expectations, there is now good evidence that there is a connection between beverage drunk and coin-tossing scores. The question is, Should the statistics student pass his statistics course?

Unfortunately, the answer is that both the statistics student and the gamblers have been misled by the regression effect. The regression effect here concerns the possible relationship between a person's initial coin-tossing score and their postdrinking coin-tossing score. Common sense dictates that there is no relationship between coin-tossing results and beverage drunk. This, then, is the regression effect seen here in its purest form. It is the regression effect when in fact there is no relationship between the two variables: the number of heads in the first toss of 100 coins and the number of heads in the second toss of 100 coins. Those who score fewer heads than average in the first toss of 100 will tend to perform about average on the second. Those who score more heads than average in the first toss of 100 will also tend to perform about average on the second.

END OPTIONAL

Correlation

We know that β, the slope of the regression line, is a method of measuring the strength of the association between the *x* values and corresponding *y* values. However, it seems desirable to have another measure that avoids the asymmetry between the variables inherent in fitting the regression line.

A statistical measure that avoids making an arbitrary distinction between the independent and dependent variable is known as the "sample correlation coefficient" (or Pearson's correlation coefficient). The sample correlation coefficient is usually denoted *r* and the equivalent true value for the population is denoted ρ (Greek letter rho), just as the sample standard deviation is denoted *s* and it estimates the true population standard deviation, denoted σ. The correlation coefficient (sample or population, *r* or ρ) takes values between –1 and +1. The further the correlation coefficient is from 0, the stronger the association between the two variables. A correlation coefficient close to +1 means that higher values of x_i are almost always associated with higher values of y_i. A correlation coefficient close to –1 means that higher values of x_i are almost always associated with lower values of y_i. A correlation coefficient of zero means that knowing the *x* value gives us no information about the corresponding *y* value.

Complicated formulas and mathematical instructions have been avoided in this book unless they give particular insight. The formula for the sample correlation coefficient is not too frightening and is perhaps a little bit enlightening and so is included for those who are interested. It is

$$\frac{\text{The average of all terms of the form } (x_i - \bar{x})(y_i - \bar{y})}{\text{std dev of } x \times \text{std dev of } y}$$

or, more precisely,

$$\frac{\dfrac{1}{n-1}\sum_{i=1}^{n}(x_i - \bar{x})(y_i - \bar{y})}{\text{std dev of } x \times \text{std dev of } y}$$

If an above-average x value is likely to have its accompanying y value also above average, we see the term $(x_i - \bar{x})(y_i - \bar{y})$ is likely to be positive. If a below-average x value is likely to have its accompanying y value also below average, we see the term $(x_i - x)(y_i - y)$ is also likely to be positive because a negative times a negative is a positive. The denominator standardizes things so that the maximum value of the sum of these terms is 1. Hence, we see that a strong tendency for the y value to be large (small) when the x value is large (small) gives a correlation coefficient close to 1. If knowing the x value had no bearing on the associated y value, the terms that are averaged are equally likely to be positive or negative and so we would obtain a sample correlation coefficient close to 0.

For example, since greater than average height tends to be associated with greater than average weight and, conversely, lower than average height tends to be associated with lower than average weight, height and weight are positively correlated: The correlation coefficient will be above 0. The coefficient would be 1 exactly if being a certain amount taller than average always meant being a proportionate amount heavier than average so that the relationship between height and weight was given exactly by the equation $w = \alpha + \beta \times h$, the equation for a straight-line relationship. Of course, weight and height are not related exactly in this way; there is some additional random scatter so that people of the same height may have different weights. As a result, the correlation coefficient for the association between height and weight is less than 1. An example of variables that are likely to be negatively correlated are number of hours of TV watched per week and exam marks. More television will often mean less time for study and this will tend to result in lower exam marks.

It should be noted that both the slope of the regression line and the correlation coefficient are valid measures of the strength of an association only on the assumption that the association of interest is in the form of a scattering about a straight line on a graph of the two variables: The variables are associated linearly. If for some unknown reason we plotted the horizontal and vertical position of a playground swing at random times, then the data we obtained would have a correlation coefficient close to zero. The plotted data would consist of points on a circular arc. There would be no tendency for high vertical values to be associated with large positive horizontal values, for high vertical values are just as likely to be associated with large negative horizontal values. There is a strong, almost perfect association between vertical and horizontal position—knowing the vertical position tells us that the horizontal position must be one of two possibilities—but this association is not linear.

The correlation coefficient r has a number of advantages compared to β, the slope of the regression line, as a measure of the association between two measurements such as height and weight. As already stated, the correlation coefficient is not affected by any arbitrary choice of one variable to be an independent variable and the other to be a dependent variable; the method of calculating the correlation coefficient treats both variables equally. Another disadvantage of β compared to r is that the value of β is affected by the units of measurement. If we measure height in meters, the slope of the line relating weight to height is 100 times steeper than if we measure height in centimeters. However, the strength of the association between height and weight is obviously not affected by our choice of measurement units. Reflecting this, the correlation coefficient is not affected by the choice of units. A third advantage for the correlation coefficient, or at least the square of its value, is that it has a nice physical interpretation: The square of the correlation coefficient is a measure of the proportion of the variability of the y values (as measured by variance) that is explained by the variation in the x values. On the other hand, the correlation coefficient has a disadvantage. Unlike β, it does not immediately slot into a formula for predicting the average value of y given a particular value of x. Both β and r are useful as measures of association between pairs of variables and each has its particular advantages.

If the slope of the regression line is zero (i.e., $\beta = 0$), it indicates that there is no linear relationship between x values and y values. As a possible relationship between the x values and y values is often a question of interest, there is theory to test the hypothesis that $\beta = 0$ (and to obtain confidence limits for β). Similarly, there is a test of the hypothesis that the correlation coefficient r is 0 (and confidence intervals can also be obtained). The tests on β and r are equivalent to each other in helping to decide whether an apparent association between two variables is "for real" or just due to chance. Both tests are valid only if we have the situation where there is a certain type of randomness. In particular, it is assumed there is an underlying straight-line relationship (the null hypothesis is that this straight line is horizontal), and for all points the vertical variability about this straight line is given by figures obtained independently from identical normal distributions.

There is a further use in statistics for the correlation coefficient r in describing the population of pairs of values on the basis of the scatter of a sample of pairs of values. This use requires more assumptions about the data and the underlying population. The term "bivariate normality" is used here, but further discussion is beyond the scope of this book.

The sample correlation coefficient (or Pearson's correlation coefficient) is defined by a formula on page 218 using the actual values of x_i and y_i. There is also a nonparametric correlation coefficient that is given the name Spearman's ρ. It is calculated by replacing the values of the x_i and y_i by their ranks. For example, if the ith person was 176 cm tall and weighed 75 kg and this person was ranked the thirtieth tallest and the nineteenth heaviest in the sample, then in place of $x_i = 176$ and $y_i = 75$ we would use the values 30 for x_i and 19 for y_i. By throwing away the exact value and only using the ranks, we are losing some information. However, hypothesis tests about the correlation coefficient depend on the assumption of an underlying normal distribution. If this assumption is not reasonable, we can use hypothesis tests on Spearman's rho.

The Principles of Hypothesis Testing Applied to Spearman's Rho

In some cases p value calculations using Spearman's rho are a good reminder of the principles underlying such calculations. One example of calculating a p value associated with Spearman's rho was given in the section on the number of possible arrangements of n objects on pages 56 and 57. We now give another example and use it to revise much of the earlier material in this book.

OPTIONAL

Say we wished to test the null hypothesis that there is no relationship between a person's athletic ability and his or her ability in statistics against the alternative hypothesis that the skills are positively correlated. Let us suppose that we gave four people tests in both athletics and in statistics and the result was that they have the same ordering in both areas (i.e., the person who came first in athletics also came first in statistics, and so on). We do not have to calculate Spearman's rho here using a formula: It must have the value 1, signifying perfect agreement in the ranking of abilities in the two areas. To determine the p value, we now ask, "How often would we obtain this perfect agreement by pure chance?" To answer this question, let us assign the letters A, B, C, and D to the people who came first, second, third, and fourth in athletics, respectively. We now ask, "What is the probability that the performance in the statistics test will also be in the order A, B, C, D?" We have four choices for the first position in statistics and for each of these four choices we then have three choices for the second position, and for each of these 4×3 combinations of choices for the first two positions we have two choices for the third position and just one for the fourth, so we have $4 \times 3 \times 2 \times 1 = 4! = 24$ combinations of choices. Under the null hypothesis there is no relationship between statistical ability and athletic ability, so all combinations of choices are equally likely. Therefore, the probability agreement in rankings at least as good as the agreement seen here—perfect agreement—is $1/24$. If (and it is a big "if") we believed that in this case it was appropriate to use the traditional benchmark p value of 0.05 or $1/20$, we would now prefer the alternative hypothesis that there is a positive association between abilities in athletics and statistics.

We should now write a paper for a learned journal reporting these findings and when the paper is published we should perhaps put out a press release telling the world at large that we have found reasonably convincing evidence that ability in statistics and athletics are positively correlated. Readers here may object on the grounds that our conclusion is based on results from just 4 people. This is not a valid objection, provided the 4 people were chosen randomly. Our conclusions here are just as valid as if we had 4,000 people and found that this large sample led to a positive correlation that under the null hypothesis gave a p value of $1/24$. The convincing evidence is not the numbers involved in the study. The convincing evidence is that it is not easy for chance alone to explain the positive correlation and we have decided that relying on a $1/24$ chance (or any chance less than 0.05) as an explanation for our results is less satisfactory than believing that there is a real correlation. The difference between a study of 4 people and a study of 4,000 people is that in the 4,000-people study a p value of $1/24$ will result from a much weaker association between abilities in athletics and statistics. Those readers who still object to the conclusions based on just 4 people should understand that the only logical source of objection is that they intuitively don't regard a p value of $1/24$ as providing reasonably convincing evidence. In other words, such objectors should logically use a benchmark p value much smaller than the conventional 0.05. This may well be reasonable given that the alternative hypothesis may not be easy to accept. However, all should understand that this 0.05 is the benchmark that is widely used as the criterion for making decisions about hypotheses in many areas of human knowledge. Clearly, it is inappropriate to use this benchmark without considering the plausibility of the hypotheses and the costs of error.

If, however, instead of the one-tail alternative hypothesis that abilities are positively correlated we had the two-tail alternative hypothesis that abilities are correlated either positively or negatively, we would not reject the null hypothesis using the benchmark p value of 0.05. This is because both the ranking we obtained—A, B, C, D—and another ranking—D, C, B, A—lead to a Spearman's rho equally far from the null hypothesis ideal Spearman's rho of 0. In other words, two of the twenty-four equally likely rankings (including the one we actually obtained) are at least as far as our ranking from the ideal under the null hypothesis. Therefore, if the null hypothesis is true and pure chance alone is operating, the probability of being this far away in the rankings from the ideal situation of Spearman's rho = 0 in this two-tail test is $2/24$ or $1/12$. Since $1/12$ is greater than 0.05 or $1/20$, we will not think that we are holding onto the null hypothesis in the face of an unreasonably small chance. We see that using four people, a two-tailed test, a benchmark *p* value of 0.05, and Spearman's rho as our test of association, it will never be possible to reject the null hypothesis. If, in fact, the null hypothesis is incorrect, we will inevitably make a Type II error. Using the term "power" as defined in Chapter 6, we can say that this experiment has no power.

END OPTIONAL

The ideas used to calculate *p* values for the null hypothesis Spearman's $\rho = 0$ in this example can in principle easily be extended. If *n* individuals each have two variables measured on them, simply calculate the Spearman's ρ for the measure of the extent to which the rankings match. Then write down all $n \times (n-1) \times (n-2) \times (n-3) \times \ldots \times 3 \times 2 \times 1 = n!$ possibilities for how the rankings could match. For each of these possibilities calculate Spearman's ρ. Under the null hypothesis each of these possibilities is equally likely: Each occurs $1/n!$ of the time. For the *p* value, simply count up the number that give a Spearman's ρ at least as big as the one obtained from the sample and divide by $n!$. This is the *p* value: It answers the question, "How easy is it for chance alone to explain results that are at least as extreme as the results in our sample?" In practice, this approach is too tedious even for a computer unless *n* is a small number. Unless *n* is small, some clever theory based on the central limit theorem and the normal distribution is used to calculate an approximate *p* value for various values of Spearman's ρ.

SUMMARY

- ANOVA is used to see if there is convincing evidence that the numbers in different groups come "from different hats." ANOVA actually starts by assuming that the numbers in all groups all come "from the same hat." It then tells us how often the means of the different groups chosen from the same hat would be at least as widely scattered as they are with our data. This can be calculated from the observed variability of the individuals within each of the groups. This is the *p* value. ANOVA assumes that we are dealing with a situation in which numbers in all the different hats are normally distributed with the same variance. If these assumptions are violated, use the Kruskal–Wallis test.

- Regression is used to assess any straight-line relationships or association between two variables measured on each of a number of individuals. If the slope of the straight line, β, is zero, then the independent variable has no effect on the dependent variable. There is a test to see if β calculated from a sample provides reasonably convincing evidence that the true β for the population is not zero. The evidence is provided in the form of a *p* value (this test indicates that there is an association between the variables in the population). Confidence intervals can be obtained for the true β for the population. Certain assumptions must be met for this test to be valid. Deciding that an association is real still leaves open the issue of causality, as in Chapter 7. The asymmetry between independent variables and dependent variables causes a difficulty known as the regression effect that reflects the fact that one variable only partly explains the other variable and chance is also involved.

- Correlation also measures the amount of linear association. Correlation, unlike the slope of the regression line, is not affected by units of measurement and does not distinguish between dependent and independent variables. However, it is not as convenient as β in obtaining a formula relating two variables. It takes values between −1 and +1, with 0 indicating no linear association. As in the case of β, the sample correlation coefficient can be used for an hypothesis test that the population correlation coefficient is zero and to obtain confidence intervals. There is a nonparametric version of the sample correlation coefficient known as Spearman's ρ. The method of finding p values testing the hypothesis that Spearman's $\rho = 0$ involves very simple mathematical ideas.

In short, use ANOVA to see if there is reasonably convincing evidence that a number of different groups have different means. Use regression or correlation to see if there is reasonably convincing evidence that a larger than average value for one variable is associated with a larger (or smaller) than average value for another variable measured on the same individual. Use regression if the aim is a formula connecting the two variables. Use correlation if a formula is not the aim and a measure of the closeness of the association in terms of a number between −1 and +1 is required.

QUESTIONS

1. The day of the week on which a number of statistics students were born was recorded along with their marks for a statistics course. The results are displayed here:

Monday	Tuesday	Wednesday	Thursday	Friday	Saturday	Sunday
58	6	21	71	39	62	81
97	19	58	86	58	31	94
43	55	23	51	26	37	65

Perform an appropriate statistical test and state with reasons whether you believe that there is a relationship between the day of birth and performance in statistics.

2. A skeptical farmer wonders whether there is really any benefit in terms of yield in the various high-yield varieties of wheat produced by agricultural researchers. Accordingly, he keeps a record over the years of the yield with various varieties. His results are as follows:

Traditional variety	2.7	1.9	3.6
Improved variety type 1	2.3	3.1	4.0
Improved variety type 2	4.4	2.7	3.5

The farmer uses a statistical package to analyze these results using ANOVA and concludes that there is no evidence that the improved varieties are better. Do you agree with the farmer? Discuss.

3. In a study of geographical temperature variation, research assistants in Townsville (latitude 19°30′), Brisbane (latitude 27°30′), Sydney (latitude 34°00′), and Melbourne (latitude 37°45′) were each asked to choose, at random, two days in May and record the maximum temperatures on those days. The results are as follows:

Townsville	27	31
Brisbane	24	18
Sydney	16	22
Melbourne	13	21

Perform an ANOVA test on these data. Comment on the null hypothesis and the test used. In light of these results, is it reasonable to believe that the maximum daytime temperatures in May in these various cities are the same?

4. The following list gives the rainfall in centimeters in Brisbane, Australia, in April of each year and the number of books imported into the United States in December of the same year. Data from the years 1995 to 2000 are given. Use an appropriate statistical method to test the null hypothesis that there is no relationship between Brisbane rainfall and U.S. book imports. Do you believe that knowing Brisbane's rainfall in April this year will assist you in predicting the book imports in the United States in December? Discuss.

Year	1995	1996	1997	1998	1999	2000
Rainfall	10.1	3.9	15.7	9.1	6.3	12.9
Books	150,060	90,210	197,300	137,980	121,100	181,600

5. In the following list the row marked x gives the prices paid by the previous owners to purchase houses in a particular town in 1990; the row marked y gives the price paid by the current owners to purchase the same houses in 2000.

| x | 33 | 37 | 38 | 39 | 41 | 36 |
| y | 67 | 73 | 74 | 75 | 97 | 84 |

Comment on the null hypothesis H_0 that there is no relationship between the purchase price paid by the current owner and the purchase price paid by the previous owner and use the results given to complete an appropriate statistical test of the null hypothesis. Do you believe that knowledge of the previous purchase price of a house will be of no use in predicting the current purchase price? Discuss.

6. Six people are questioned by a researcher and are ranked according to the amount of daily exercise that they undertake. The same six people also have a blood cholesterol test. It turns out that the person who undertakes the most exercise also has the lowest cholesterol level. The person who undertakes the second greatest amount of exercise also has the second lowest cholesterol level, and so on, so that the person with the least exercise has the highest cholesterol. Say this experiment was repeated many times. Find how often on average this inverse matching of rankings of cholesterol level and exercise would occur if in fact there were no connections between cholesterol and exercise. If the experiment is in fact carried out only once and gives the results described, would you be reasonably convinced that there is an inverse association between exercise and cholesterol levels?

7. A researcher looking for evidence of climate change notes the ten localities in America with the highest rainfall totals for the twenty-four hours ending at midnight, January 2–3, 2001. The following year he checks the rainfall in the same ten localities over the same twenty-four-hour period. He finds that in all cases the rainfall in these localities on January 2, 2002, is less than it was on January 2, 2001. The researcher argues that even a simple sign test of these results gives a *p* value of 1/1,024 (one-tail), providing strong evidence against the null hypothesis that America's rainfall is not changing. Discuss any flaws that you may see in this researcher's experiment and conclusion.

8. The rabbit populations of 100 fields were estimated and the 9 fields with the highest estimated rabbit populations were selected for a rabbit inhibition experiment. The inhibition experiment consisted of erecting gruesome pictures in the middle of each field portraying farmers killing rabbits. After three years these nine fields were revisited and rabbit numbers were again assessed. In all nine cases the rabbit population had fallen. Assuming the experimental method is valid, perform an appropriate test to determine the probability of an improvement in nine out of nine cases being achieved if in fact the gruesome pictures had no effect on the rabbits. Are you reasonably convinced that gruesome pictures inhibit the rabbit population? What major flaws are there in the experimental method here?

NOTE

1. An *f* ratio of 1.0 corresponds to the ideal situation, when the null hypothesis is true, of both estimates of individual variability being equal. It may therefore be thought that the *f* ratio will exceed 1.0 exactly half the time; that is, the associated *p* value should be 0.50. However, asymmetry in the *f* distribution means that there is a less than 50 percent chance of exceeding the value of 1.0.

CHAPTER 10

Miscellaneous Topics

FINITE POPULATION CORRECTION

Often in statistics we are dealing with a population that is effectively infinite and we want to learn about the population by taking a sample comprising a limited number of values drawn from this virtually infinite population. For example, if we want to learn something about the effect of a disease and we examine some people with the disease, we are generally sampling an almost infinitesimal proportion of the population of all those people who now have or might ever have the disease. If we want to learn something about the pollution levels in a town's air over time, the amount of air we sample would be an almost infinitesimal proportion of the air that circulates over the town during a period of time. The sample is useful, not directly because it gives us knowledge about some appreciable fraction of the population, but because it gives us probabilistic ideas about the entire population.

Sometimes, however, we will deal with a population that is not particularly large in comparison to the size of our sample. For example, we might be interested in environmental attitudes of the mayors of all the local authorities in Australia. The question then arises, Why not do a complete census? With a complete census there is no need for statistics in the sense that there is no uncertainty: Confidence intervals and hypothesis tests regarding the true levels of environmental awareness, should all the mayors be questioned, are irrelevant, as they all have been questioned.

Sometimes we will do a complete census, but often we won't, for two reasons. The first reason is that it may be too expensive. The second reason is that with limited resources we may get more accuracy by concentrating our resources on a sample rather than spreading them out over the whole population. Returning to the example of the environmental attitudes of the mayors, if we were to conduct a census of all of them, our resources might limit us to a postal survey that might then be filled out in haste by the mayors' aides. If, however, we were to concentrate our resources on a sample, we might be able to get much more thoughtful responses from all the mayors in the sample using face-to-face interviews.

However, if our sample consists of an appreciable proportion of the entire population, our statistical analysis has to be modified. The modification reflects the fact that our sample not only gives us a probabilistic idea of what we would expect from the rest of the population, but it also gives us precise information about an appreciable proportion of the whole population. Mathematical theory we will not cover shows that the appropriate modification is quite straightforward. Our ideas on how uncertain our estimate of the population mean is—the standard error of the mean—have to be reduced by multiplying the standard error of the mean by $\sqrt{1-f}$, where f is the proportion of the population in the sample. This adjustment is known as the finite population correction. The statistical analysis then proceeds as usual. If we are using a computer program in our analysis, the program may display the standard error of the mean so that we can factor in a finite population correction manually, if appropriate. For example, say we sampled a random sample of ten of the mayors of Australia's twenty largest cities and found the following results on some numerical environmental awareness scale: 5, 6, 9, 4, 8, 10, 7, 3, 1, 2. We will assume here that it reasonable to analyze these figures as though they came from a normal distribution and so use the methods described in Chapter 5. The mean score is 5.5. The sample standard deviation is 3.027. The standard error of the mean without using the finite population correction is $3.027/\sqrt{10} = 0.957$, but with the finite population correction it is $0.957 \times \sqrt{1 - {}^{10}/_{20}} = 0.677$. If we wanted 95 percent confidence intervals for the true mean of the scores of the mayors on this environmental awareness scale, it would be $5.5 \pm 0.677 \times 2.262$ where ± 2.262 is the range of values from the t_9 distribution that contains 95 percent of values. In other words, the 95 percent confidence interval would be (4.0, 7.0). If we had not used the finite population correction we would have (3.3, 7.7). The latter calculation would ignore the fact that not only do we have an impression of what all the mayors are like from interviewing ten of them, but we also have certain knowledge about what half of them are like. In general, the $(100 - \alpha)$ percent confidence interval for the mean as a result of a survey from a finite population is given by

$$\bar{x} \pm \tau \frac{s}{\sqrt{n}}\sqrt{(1-f)}$$

where n is the number surveyed, f is the proportion of the finite population surveyed, and τ is the figure from the t_{n-1} distribution such that $(100 - \alpha)$ percent of the values from this distribution are in the range $-\tau$ to $+\tau$ (it is assumed here that the values that occur in this finite population are values that are chosen from a normal distribution).

Sometimes there is a philosophical difficulty here. If we want to know about the actual mayors of large Australian cities, we would use the finite population correction factor as in the preceding paragraph. However, if we were thinking of these mayors as a representative sample of all the mayors who could ever exist given the same social circumstances as exist in Australia, we would regard our population as infinite and not use the population correction factor. In the same way, if we have a statistics class with male and female students, we may find that the average mark of the females on the examination is 1 percent better than the average mark of the males. We could then ask the question, "Is this due to chance?" From one point of view, this is a meaningless question. Our sample is the population. We know precisely the mark of everyone in the class. Knowing these marks, there is no chance that these results for the population, the class, could be anything other than the results that we have in front of us. We can say, dogmatically, that in this class we are absolutely sure that, on these marks, women are better on average than men. From another point of view, we can think of this class as just a sample of all the billions of men and women who could potentially enroll in a class such as ours. Assuming the class is not large and that there is considerable individual variation in marks and noting the small difference in the average mark, after a statistical analysis we could conclude, "There is no (convincing) evidence that women are better than men."

DETERMINING THE NUMBER OF SAMPLE VALUES REQUIRED

The amount of data required depends on how accurate we want our estimate to be and how variable the individual values are. We will consider two cases: estimating a proportion and estimating the mean of values that are normally distributed.

Estimating a Proportion

Say we wanted to find out the proportion of people intending to vote for a particular political party. This involves the binomial random variable (Chapter 4). When the sample is of considerable size (e.g., twenty or more), the binomial distribution "looks" like a normal distribution, as it is the result of adding a considerable number of chances (this is the central limit theorem; see Chapter 5). If the true proportion in the population is θ, then the expected (anticipated average) number of successes in a sample of size n is $n\theta$ and the standard deviation is $\sqrt{n\theta\phi}$ (see Chapter 4 and recall that in the notation used there $\phi = 1 - \theta$). Then, by the central limit theorem, the actual number of successes in the sample will be approximately a value chosen from a normal distribution centered on $n\theta$ and with standard deviation $\sqrt{n\theta\phi}$. The proportion of successes in the sample will be $1/n$th of this, so the proportion of successes in the sample will be chosen from a normal distribution centered on θ and with standard deviation

$$\frac{\sqrt{n\theta\phi}}{n}$$

(this can be rewritten as $\sqrt{\dfrac{\theta\phi}{n}}$).

Therefore, 95 percent of the time the proportion θ that will be obtained will be in the range

$$\left[\theta - 1.96\left(\sqrt{\frac{\theta\phi}{n}}\right), \theta + 1.96\left(\sqrt{\frac{\theta\phi}{n}}\right)\right].$$

We see that if we want a 95 percent chance of being no further than 1 percent away from the true value of θ, we should take n so that

$$1.96\left(\sqrt{\frac{\theta\phi}{n}}\right) = 1\% = 0.01.$$

Rearranging this equation gives

$$n = \theta\phi\left(\frac{1.96}{0.01}\right)^2.$$

Now $\phi = 1 - \theta$, and high school algebra shows that the biggest value that $\theta(1 - \theta)$ can take is $\frac{1}{4}$ (this value occurs when $\theta = \frac{1}{2}$). Approximating 1.96 by 2 and taking the largest possible value of $\theta\phi$ of $\frac{1}{4}$ shows that if we take $n = \frac{1}{4} \times (2/0.01)^2 = 10,000$ we will have at least a 95 percent chance of obtaining a proportion that is within 1 percent of the true proportion θ. Public-opinion surveys are often done using $n = 500$, not 10,000. Calculations using this theory show that these surveys can have about a 1 in 20 chance of being inaccurate by more than 4 percent.

The Number of Values Required for Estimation of the Mean of a Continuous Variable to within a Given Accuracy

The formula for the amount of data required in this situation is worked out using similar reasoning. This time, however, we have to have some preliminary estimate of the variability in order to make an estimate of n. We assume that we have an estimate s of the standard deviation. The formula, by similar reasoning to that shown earlier, turns out to be

$$n = \left(\frac{us}{d}\right)^2,$$

where we are prepared to be in error by an amount of d or more with a chance of α, and u is the value from a t distribution such that the chance of being above u is $\alpha/2$. If n turns out to be reasonably large (e.g., greater than 20), the standard normal distribution usually is used, as it is a good approximation to the corresponding t distribution. For example, if we had a variable for which preliminary information indicated that the standard deviation was 10.0 and we wanted to be 99 percent sure that our estimate of the mean was within 2.0 units of the true value, we would need

$$n = \left(\frac{2.57 \times 10.0}{2.0}\right)^2 \approx 165.$$

The figure 2.57 is used here because the range ±2.57 from the standard normal distribution contains 99 percent of values.

Our considerations here can lead to the issues raised in dealing with confidence intervals. If, with $s = 10$, we are about to examine a sample of 165, then there will be a 99 percent chance that the mean of the sample will be within 2.0 units of the mean. However, once we have obtained a particular sample mean we cannot generally say there is a 99 percent chance that the true population mean μ will lie within 2.0 units of the sample mean. Our 99 percent confidence interval will, however, be the sample mean ±2.0 units. For example, if we are dealing with the heights of women, although 99 percent of samples of 165 women may give sample means in the range 170 to 174 cm, by a very long coincidence we may have obtained a sample with a mean of 150 cm. We have prior knowledge about the likely average height of women and so it would not be correct for us to believe on the basis of our sample that there is a 99 percent chance that the average woman is between 148 and 152 cm. Ideas about where the sample mean is likely to be knowing the population mean cannot generally be inverted to ideas about where the population mean is likely to be knowing a sample mean (see the material on confidence intervals in Chapter 6 for more explanation of this issue).

TOPICS COVERED IN
MORE ADVANCED STATISTICS TEXTS

The topics covered in this text give the reader sufficient knowledge to apply statistics to most straightforward situations where only one or two measurements are made on each individual. Hopefully, this text, with its emphasis on understanding the philosophy, will enable the reader to apply statistics with common sense in such situations.

However, there are many parts of the subject of statistics that have not been covered and it seems appropriate in this last chapter to give an indication of the scope of the subject. This will be given in the form of the following list of randomly selected topics in random order:

- We have covered the common tests appropriate to certain combinations of sources of data and types of data as described at the end of Chapter 5 and expanded at the end of this chapter. However, there are many more tests applicable in circumstances we haven't considered. For example, we haven't considered the situation where there is a measure on each individual performed after each of a number of different interventions where that number is more than one.

- Mathematical statistics is a large subject. Among many other things it delves into questions of being precise about what we mean when we say things like "s as defined in Chapter 2 is a "good" estimate of σ." What are the mathematical properties of a "good" or "best" method of estimating a parameter value? Are there mathematical methods for finding the "best" method of estimation?

- Throughout this text the need to incorporate common sense into statistics has been emphasized. The reader has been urged to do this by modifying their benchmark p value according to circumstances. There are also mathematical ways of incorporating common sense into statistics. The subject of Bayesian statistics is one of a number of mathematical approaches to incorporating common sense into statistics.

- Decision theory is a further attempt to refine the use of statistics. It is a method of using objective and subjective information about probabilities and explicitly taking into account the costs of errors.

- The previous section gave some information on calculating how many sample values would be needed for a certain amount of accuracy in two simple situations. There is a lot more to the topic of finding the sample numbers necessary for a statistical analysis to have some required power. A related topic is the topic of stratified sampling. If we wanted an estimate of total amount of soil lost to erosion in Australia each year, positioning test areas at randomly chosen spots throughout Australia would not be optimal. To improve accuracy, we would be best off focusing more sampling effort on geographic regions where erosion levels were more variable.

- Being as economical as possible in terms of number of subjects in an experiment is particularly important when the decision about the best treatment is a matter of life and death. There are special methods known as sequential analysis for continually checking the data to decide when sufficient people have undergone the experimental treatment for a decision about it to be made.

- Survival analysis is a related area. In medicine, the final endpoint for many studies is death. However, as people, even sick ones, often live a very long time, we would often have to wait a very long time before everybody in a study died and we had complete results to use in comparing the benefits of different methods of delaying death. Using incomplete results when only some people have died is the subject of survival analysis.

- The practicalities of sampling, particularly sampling humans, is another large topic. For example, phone surveys don't represent people in households without a phone, but, less obviously, they underrepresent those in large households with only one phone per household (see the answer to question 2 of Chapter 2).

- Most of this book has dealt with samples in which the individuals have been chosen at random. In spatial statistics, though, the equivalent to our individuals are points we choose to sample in space. Points in space are not independent: They are related according to how physically close they are. Spatial statistics is an important component of environmental science and of geology. It is required, for example, in order to use limited information to draw maps of pollution levels, assess the population of endangered species, and assess the amount of ore in a mine.

- Just as points can't be independent in space, they can't be independent in time. Special statistical methods are required for analysis of fluctuating data through time. This is the subject of time series. Physicists looking at sunspots, meteorologists looking at weather patterns, and economists looking at fluctuations in the capitalist economies are all interested in time series.

- In biology and medicine we are often interested in a number of factors that may all be operating simultaneously in one individual to have an effect on what is being measured. We may want to know if the effect of a new drug taken for blood pressure is affected by gender, age, preexisting blood pressure level, dietary salt intake, and coprescription of certain other drugs. Furthermore, we may want to compare how the new drug and the standard drug interacts with these factors. Teasing out answers to such questions constitutes the majority of many second-level applied statistics courses for biologists. These answers come under headings such as advanced regression and multiway ANOVA or ANOVA models. These topics, in turn, are derived from a branch of mathematical statistics known as generalized linear models.

- Often more than one or two measurements are made on an individual. Particularly in psychology, vast numbers of measurements are made on each individual in the form of responses to a questionnaire containing hundreds of questions. In biology and medicine as well, it is common to take many measurements of various aspects of each individual. Making sense of all this information is the subject of multivariate analysis. Multivariate analysis includes a number of topics. For instance, principal component analysis, factor analysis, and cluster analysis can deal with condensing the mass of information from psychology questionnaires into a more manageable form. These topics give methods of answering questions about whether people's personalities tend to fall into a limited number of types or about how much of the variation between people can be summarized by, say, three figures (for example, the three figures might give a measure of intelligence, a position on a scale of introversion–extroversion, and a position on a scale of conservatism–radicalism). There are also many other uses for these techniques. In medicine, the topic of discriminant analysis is used to find a method of combining a number of indirect measures to obtain a score that is best able to discriminate between the presence or absence of a serious disease. This can avoid the need for expensive or dangerous operative treatment to decide the issue beyond doubt. There are many other topics within the area of multivariate analysis and many other uses for this branch of statistics.

SUMMARY

- Occasionally, our sample constitutes an appreciable proportion of the population of interest. In such cases, confidence intervals for means have to be reduced in width by a factor of $\sqrt{1-f}$, reflecting the fact that we are certain in our knowledge of an appreciable proportion of the population.

- We can determine the number of measurements required for a statistical test to have a certain level of accuracy.

- This text covers most straightforward statistical tests, but the scope of advanced statistics is huge.

SUMMARY OF STATISTICAL TESTS

All the statistical tests covered in this book are designed to help answer the question, "Are the differences we see 'for real' or are they just the result of chance?" The results from the statistical tests are not a direct answer to this question, but instead tell us how easy it would be for chance to explain the results.

The term "differences" in this context has several shades of meaning:

1. We may be interested in the possibility of a difference of an appreciable size caused by a definite intervention or membership of a definite group. We have covered a number of such tests. Different tests apply in different situations, depending on the source of data and the type of data. These tests were shown at the end of Chapter 5, and the list is repeated here with the addition of tests covered later:

Source of data	Dichotomous (e.g., better or worse) data	Numerical but not necessarily normal data	Numerical and normal data
Two related measures (e.g., measures before and after an intervention on the same individual; measures on one twin who had an intervention and the other twin who didn't)	Sign test	Wilcoxon signed rank test	Paired samples *t* test
A single measure on two unrelated samples (e.g., measuring the same quantity on men and women)	Fisher's exact test	Mann–Whitney test	Independent samples *t* test
A single measure on more than two unrelated samples	χ^2 test of association (applies to nominal not just dichotomous outcomes)	Kruskal–Wallis test	ANOVA

2. We may be interested in shades of difference where there is no distinct group as such but individual measures are at different points along a continuous range of values. In this case, we usually use the word "association" and ask if variation in one measure is associated with variation in another measure. Here we use correlation and regression. Valid *p* values can be calculated for the measures that we obtain from correlation and regression when the association is linear and the appropriate test is used. There are several types of data:

- Both measures are numerical but not normal.

  Test: Spearman's ρ (rho)

- At least one of the measures is normally distributed as the dependent variable scattered about a regression line.

  Test: to see if the slope of the regression line (β) or Pearson's correlation coefficient (*r*) is zero

- The two measures have a bivariate normal distribution.

  Test: (Pearson's) correlation coefficient

3. We may be interested in the possibility of a difference between the population from which we have drawn our sample and some theoretical distribution.

The relevant tests that we have covered include the following:

- the use of the binomial random variable to test an hypothesis about the value of the parameter θ or the proportion in the population, based on knowledge about a sample.

- the use of the Poisson random variable to test an hypothesis about the value of the parameter λ or the average rate at which something happens.

- the use of the *z* test to test whether a single value or the average of a number of values comes from a normal distribution with μ and σ specified.

- the use of the single sample *t* test to test whether the average of a number of values comes from a normal distribution with just μ specified.

- the use of the Komogoroff–Smirnoff test, mentioned in this book as a test of the data fitting a normal distribution, but with wider applicability.

- the use of the chi-square goodness of fit test to test whether the proportions of the data in various categories are a reasonable match to those expected by the theoretical distribution.

Most of the tests under (1) and (2) can also be used to define confidence intervals. Confidence intervals are a way of using the measures on the sample to obtain information on the likely position of some unknown parameter that describes the population or describes the average difference made by some treatment. Again, note that a confidence interval does not directly tell us that the unknown parameter is in a certain interval with a certain probability. Instead, the confidence interval tells us that if the parameter was in this interval it could "easily" have given the underlying data. The term "easily" refers to calculations with a given value of the parameter, giving a *p* value for the data larger than the benchmark *p* value.

QUESTIONS

1. Say that the entire adult population of northern hairy-nose wombats consists of 180 individuals and that we survey a randomly selected sample of 100 of these individuals and measure their weights. If the results are that the mean is 27 kg and the standard deviation is 2 kg, find the 95 percent confidence interval for the mean weight of adult northern hairy-nosed wombats.

2. On a test in a statistics class the mean mark was 69.71 and the standard deviation was 11.65. However, only twenty-one of the twenty-two students now enrolled took the test, as one student was unable to take it because of ill health.

   a. Assume this one student has been attending class as much as the other twenty-one students until the time of the test. Find the 95 percent confidence interval for the mean mark that would have been obtained in the first test had all twenty-two students been able to take it.

   b. The absent student's health deteriorates further and she withdraws from the class. What now is the 95 percent confidence interval for the mean mark of the class?

3. Consider a survey on an issue in which the American population is thought to be approximately evenly divided.

   a. How many people would we need to survey for us to have a 99 percent chance of obtaining a value that differs from the true percentage by no more than 1 percentage point?

   b. After taking our survey using an appropriately chosen representative sample and with the appropriate numbers obtained from part a, we find that 40 percent of the sample is positive about the issue under consideration. Does this mean that we can be 99 percent certain that the true proportion of the whole of the American population who are positive about this issue is between 39 and 41 percent? Discuss.

4. For each of the following scenarios state the most appropriate statistical test.

   a. People are classified into two groups depending on whether they were abused as children and the recorded outcome is whether they have had a conviction for theft.

   b. Groups of people in four different industries are selected at random and their blood pressures are measured, as the investigators are interested in a possible association between blood pressure and type of workplace.

   c. It is thought that the position of a person's surname in the alphabet may, as a result of childhood experiences of waiting for names to be announced in alphabetical lists, lead to personality differences. Accordingly, a psychological test that gives a numerical score on an introversion–extroversion scale is administered to a group of people. About half the people in the group have surnames starting with the first four letters of the alphabet and the remainder have surnames starting with the last four letters of the alphabet. The introversion–extroversion score of all these people is assessed. Assume that inspection of the figures suggests that it is reasonable to believe that the data come from a normal distribution.

d. Consider the same scenario as in c, with names starting with letters at different ends of the alphabet and measurement of an introversion–extroversion score. What test should we use to assess the results if inspection of the figures suggests that it is not reasonable to believe that the data come from a normal distribution?

e. A piped-music system is installed in a hospital for long-stay patients and the recorded outcome is whether patients felt better or worse on a day when they had access to the music than on a day when they didn't have access to the music.

f. A piped-music system is installed in a gym and the recorded outcomes are the amount of weight each member of the gym could lift on a day without piped music and the amount of weight each member could lift on a day with piped music. Assume that inspection of the figures suggests that it is reasonable to believe that the data come from a normal distribution.

g. Consider the same scenario with the gym and piped music as in f, but this time assume that inspection of the figures suggests that it is not reasonable to believe that the data come from a normal distribution.

h. People are classified into two groups, depending on whether they grew up in a rural or urban setting, and the recorded outcome is whether they are vegetarian.

i. Groups of women, all age twenty, who adhere to four different religions are selected at random and their time for a 100-meter sprint race is recorded, as the investigators are interested in a possible association between athletic performance and religion.

j. We are interested in the possibility of an association between religion and occupational group.

k. People of varying incomes are selected and their IQs are measured, as the investigators are interested in a possible association between income and IQ.

Appendix: Table of the Standard Normal Distribution

z	0	1	2	3	4	5	6	7	8	9
0.0	5000	5040	5080	5120	5160	5199	5239	5279	5319	5359
0.1	5398	5438	5478	5517	5557	5596	5636	5675	5714	5753
0.2	5793	5832	5871	5910	5948	5987	6026	6064	6103	6141
0.3	6179	6217	6255	6293	6331	6368	6406	6443	6480	6517
0.4	6554	6591	6628	6664	6700	6736	6772	6808	6844	6879
0.5	6915	6950	6985	7019	7054	7088	7123	7157	7190	7224
0.6	7257	7291	7324	7357	7389	7422	7454	7486	7517	7549
0.7	7580	7611	7642	7673	7703	7734	7764	7793	7823	7852
0.8	7881	7910	7939	7967	7995	8023	8051	8078	8106	8133
0.9	8159	8186	8212	8238	8264	8289	8315	8340	8365	8389
1.0	8413	8438	8461	8485	8508	8531	8554	8577	8599	8621
1.1	8643	8665	8686	8708	8729	8749	8770	8790	8810	8830
1.2	8849	8869	8888	8907	8925	8943	8962	8980	8997	9015
1.3	9032	9049	9066	9082	9099	9115	9131	9147	9162	9177
1.4	9192	9207	9222	9236	9251	9265	9279	9292	9306	9319
1.5	9332	9345	9357	9370	9382	9394	9406	9418	9429	9441
1.6	9452	9463	9474	9484	9495	9505	9515	9525	9535	9545
1.7	9554	9564	9573	9582	9591	9599	9608	9616	9625	9633
1.8	9641	9649	9656	9664	9671	9678	9686	9693	9699	9706
1.9	9713	9719	9726	9732	9738	9744	9750	9756	9761	9767
2.0	9772	9778	9783	9788	9793	9798	9803	9808	9812	9817
2.1	9821	9826	9830	9834	9838	9842	9846	9850	9854	9857
2.2	9861	9864	9868	9871	9875	9878	9881	9884	9887	9890
2.3	9893	9896	9898	9901	9904	9906	9909	9911	9913	9916
2.4	9918	9920	9922	9925	9927	9929	9931	9932	9934	9936
2.5	9938	9940	9941	9943	9945	9946	9948	9949	9951	9952
2.6	9953	9955	9956	9957	9959	9960	9961	9962	9963	9964
2.7	9965	9966	9967	9968	9969	9970	9971	9972	9973	9974
2.8	9974	9975	9976	9977	9977	9978	9979	9979	9980	9981
2.9	9981	9982	9982	9983	9984	9984	9985	9985	9986	9986
3.0	9987	9987	9987	9988	9988	9989	9989	9989	9990	9990
3.1	9990	9991	9991	9991	9992	9992	9992	9992	9993	9993
3.2	9993	9993	9994	9994	9994	9994	9994	9995	9995	9995
3.3	9995	9995	9995	9996	9996	9996	9996	9996	9996	9997
3.4	9997	9997	9997	9997	9997	9997	9997	9997	9997	9998

Note: The decimal point at the start of each probability is suppressed for readability. Probabilities corresponding to larger values of z can be calculated using the following approximation:

$$\text{probability of a value greater than } z \approx \frac{1}{\sqrt{2\pi}} \frac{1}{z} e^{-z^2/2}$$

Answers

CHAPTER 1

1. There are innumerable possible examples. One used later in this book concerns attention to hygiene and a decrease in the chance of infection.

2. The roulette has no way of "remembering" what it did the last 500 times. Therefore, the only relevant details are that it is fair and so has thirty-eight equally likely places where it can land. The answer, then, is that the probability of landing on a "36" is 1/38, so the probability of losing is 37/38.

3. There is no answer that is right for everyone. This is the point: Most decision making cannot be entirely objective. Someone who believes that clairvoyance exists and is not uncommon may well be convinced by a correct guess of a number between 1 and 10, whereas the skeptic referred to in the text would want a correct guess of a number between 1 and 1,000,000,000.

4. If you are being logically consistent, your answer to question 4 should be about the square root of the number you gave as the answer to question 3 (note that m is the square root of n if $m \times m = n$, so, for example, 10 is the square root of 100 because $10 \times 10 = 100$). For example, if you think that a correct response would be just sufficient to convince you if there was only 1 chance in 100 of getting the right answer by guessing alone, then two correct guesses of a number between 1 and 10 should be just sufficient to convince you. To see this, reason that for each of the ten possibilities for the first guess there are ten possibilities for the second guess, so with two guesses of a number between 1 and 10 there are 10×10 possible combinations for the two guesses. Another way of seeing this is to change the problem slightly so that the numbers to be guessed can be any of the ten numbers 0 to 9. Then guessing any of the 100 two-digit numbers 00 (or 0) to 99 is the same as using the first guess to guess the first digit of the two-digit number and the second guess to guess the second digit of the two-digit number.

CHAPTER 2

2. There are several answers:

- How were the names "chosen at random"? If the person performing the poll chose names "at random" by eye from the phone book, he or she might unconsciously tend to choose surnames that are likely to belong to his or her own ethnic group. Other ethnic groups may tend to have a different opinion, and they will not be fairly represented.

- Opinions of people too poor to own phones will not be represented. The same is true of people who choose to have "unlisted" numbers. Both these groups may tend to have different opinions from those who have listed phone numbers.

- People who live in crowded households will be underrepresented by this polling method. To see this, imagine a village of 100 people where 10 people live on their own in single households and all the others live in households of 9 people. If every household in the village has a phone and one person from each household answers the opinion-poll phone call, half the opinions obtained from the twenty phones in the village will be from people in single households whereas only 10 percent of people in the village live in such households.

- If those who don't answer the phone are not pursued, then the opinion poll will tend to underrepresent the opinions of people who have lives that take them out of reach of the telephone and conversely may tend to overrepresent the opinions of people, such as elderly retirees or mothers of young children, who may tend to spend more time at home.

- The opinions of those who refuse to give their opinions to opinion pollsters will not be represented.

CHAPTER 3

1. (a) (i) 80/140 or 4/7   (ii) 60/140   (iii) 30/140   (iv) 110/140

   (b) 110/140 = 80/140 + 60/140 − 30/140

   (c) neither

2. (a) 42/10,000 or 0.42%   (b) 58%   (c) 3% + 14% − 0.42% = 16.58%   (d) 93.7%

3. $1/2 \times 3/4 + 1/3 \times 1/4 = 11/24 \approx 45.83\%$

4. 4.475%

5. (a) 0.1064 or 10.64%   (b) 0.0288/0.1064 = 0.2707 or 27.07%

   (c) 0.916 or 91.6% and 0.912/0.916 = 0.9956 or 99.56%

6. $2/3 \approx 66.67\%$

7. Probability land, 0.3 or 30%; probability sea, 70%

8. (a) 5! = 120   (b) 20   (c) 10   (d) 0.1

9. (a) 1/252 or about 0.003968 or 0.3968%

   (b) As in all decision making informed by statistics, there is no absolutely correct answer. However, to me it would seem more reasonable to believe that the explanation for the selection of all girls is that the teacher has a sex bias rather than to believe that the teacher is unbiased and the selection is explained by a 1/252 chance coming off.

(c) 2/252

10. (a) $1/252 \approx 0.00398$

(b) The comment given as answer to 9 (b) applies. My benchmark p value (i.e., the value just sufficient to convince me that the association is real and not just due to chance) would be 1.0 for (i), 0.000000001 or smaller for (ii), and 0.05 for (iii). This is because I would reason that (i) riding without lights at night is bound to be more risky, (ii) it is almost impossible to imagine how birthdate could affect liability for accident, and (iii) sports enthusiasts may have more skill but also may take more risks when commuting by bike. On the other hand, neither of these factors may apply to their commuting. In the absence of moderately convincing evidence to the contrary, I would not want to make a decision that sports enthusiasm makes a difference to risk. Accordingly, my answers to (b) (i), (ii), and (iii) are yes, no, and yes, respectively.

(c) p value $1/6 \approx 0.167$. Accordingly, my answers to (i), (ii), and (iii) are yes, no, and no, respectively.

CHAPTER 4

1. (a) (i) $^1/_{1,024}$ (ii) $^{10}/_{1,024}$ (iii) $^{45}/_{1,024}$ (iv) $^{120}/_{1,024}$ (v) $^{210}/_{1,024}$ (vi) $^{252}/_{1,024}$

(b) It seems reasonable to me here to use the conventional benchmark p value of 0.05. This choice of benchmark p value might be reasonable for a doctor who doesn't want to have to store an extra fact in his or her head regarding the relative effectiveness of drugs without moderately convincing evidence. Others are entitled to a different opinion: An asthmatic may want to choose the drug favored by even the weakest evidence and so may use a benchmark p value close to 1. An asthmatic with this opinion would want the drug favored by even a slim majority. A two-tail test is reasonable (either drug could be superior). With these considerations, there would have to be 1 or fewer (i.e., 1 or 0) preferring fenoterol (or preferring salbutamol) to convince a doctor whose benchmark p value is 0.05. The p value for this result is $2 \times (1/1,024 + 10/1,024)$ 0.0215. A result of two or fewer favoring either drug would occur more than 5 percent of the time, as $p = 2 \times (^1/_{1,024} + ^{10}/_{1,024} + ^{45}/_{1,024}) \approx 0.109$.

2. (a) p (one-tail test) $= ^{11}/_{1,024} \approx 0.0107$. A one-tail test may be appropriate, since presumably there is a preexisting sentiment that if the machines do anything for sleep, they will enhance it. This p value could well be convincing to some people, but since I know of no medical mechanism by which the machine could work, I would remain dubious and I would require a more stringent benchmark p value of say 0.001 to convince me.

(b) As for part (a), the "ideal" treatment of the "don't knows" is debatable; many think that it is okay to simply ignore them.

(c) Ignoring the "don't knows," the computer gives a two-tail p value of 0.000425. The one-tail p value is half of this. Given my benchmark p value of 0.001, I am convinced. Note this question illustrates a major advantage of using p values rather than the actual numbers to assess the evidence: The strength of the evidence of "1 out of 10" could not be compared with the strength of the evidence provided by "65 out of 95" without a *p* value approach.

3. $p = 0.0118$ (two-tail). I would regard this as convincing evidence for more men with broken legs.

4. $p < 0.000001$ (two-tail). Convincing evidence.

5. (a) $^{28}C_0(\frac{3}{4})^{28}(\frac{1}{4})^0 + {}^{28}C_1(\frac{3}{4})^{27}(\frac{1}{4})^1 + {}^{28}C_2(\frac{3}{4})^{26}(\frac{1}{4})^2 + {}^{28}C_3(\frac{3}{4})^{25}(\frac{1}{4})^3 = 4689 \times (\frac{3}{4})^{25} \times (\frac{1}{4})^3 \approx 0.05514$

    (b) $(\frac{3}{4})^7 \approx 0.1335$

6. (a) Mann–Whitney two-tail *p* value of 0.29828 (exact calculation using pds computer program; 0.27647 using approximate calculation on computer). While teachers are unlikely to be of exactly the same effectiveness, it would seeem unfair to declare one teacher superior with evidence that could be explained by chance this easily. I would use the conventional 0.05 benchmark *p* value here.

(b) No *p* value calculation is possible in the classroom situation, as the performances of the students will not be independent. For example, one very disruptive student in one teacher's class could affect the performance of the whole class.

7. Wilcoxon signed rank two-tail *p* value of 14/1,024 (exact calculation using pds) $\approx$ 0.01367. Reasonably convincing evidence for me.

8. 0.0498, 0.149, 0.224, 0.224, 0.168, 1.81×10^{-8}

9. 0.5 cases per three months; exactly four cases, 0.00158; four or more cases, 0.00175. Knowing that new factors in the environment can trigger disease, it seems reasonable to use a benchmark *p* value that on the one hand does not demand an extreme coincidence before conceding a new environmental factor is operating but on the other hand does not provoke unnecessary alarm when events occur that could easily be due to coincidence. Perhaps the traditional 0.05 benchmark *p* value would then be appropriate. The hematologist using this traditional benchmark *p* value would conclude that a new environmental factor is responsible for these cases, as the *p* value here is 0.00175.

CHAPTER 5

1. 115.76 megawatts. In pds, click on "Statistical functions" and "Normal distribution" and enter 0.995 under the heading "Probability $Z < z$." Click "OK." This gives 2.576 standard deviations. Since each standard deviation is 10, the answer is $90 + 10 \times 2.576$.

2. (a) 0.008198. The standard error for a group of sixty-four is $^{10}/_8$ or 1.25, so 57 is $^3/_{1.25}$ or 2.4 standard deviations below the mean of 60. In pds, click on "Statistical functions" and "Normal distribution" and enter -2.4 under the heading "z value." Click "OK" to give the answer.

    (b) No calculation is possible, as students are not selected independently of each other.

3. (a) (i) zero; no one will be absolutely exactly any specified value.   (ii) 0.3183
    (iii) 0.0345

(b) 0.0345

(c) $p = 0.0027$ (two-tail). This is sufficiently convincing evidence for me, as it is quite plausible that the environments of different countries will have effects on physiological measurements.

4. A paired sample t test may be reasonable here, particularly as the standard deviation of the differences is small (3.817) compared to the speeds we are dealing with. This test gives $p = 0.0465$. The Wilcoxon signed rank test, using less information, gives $p = 0.1563$. It seems reasonable to use the result from the paired sample t test and the traditional benchmark p value here and so conclude that there is reasonably convincing evidence that brand B, on which our sample of cyclists travel faster by an average of 4.1, is superior.

5. (a) It is not reasonable to perform a test based on the normal distribution on these figures. In each group many of the figures are less than 1 standard deviation above the minimum possible value of 0. The assumption of a normal distribution would be a distortion of the reality and the p value should not be believed.

(b) independent samples t test p value 0.0620 and Mann–Whitney test p value 0.0295. Both these values assume a two-tail test, whereas a one-tail test seems more reasonable in that one would expect that any difference would act in favor of one group (the more expensive bikes).

(c) In the absence of any strong feelings, it seems reasonable to use the conventional benchmark p value of 0.05 (it may be argued that this is too restrictive in that one would expect a priori expensive bikes to be better, but it could also be that they are more delicate). Using the traditional benchmark and half the p value obtained from the two-tail Mann–Whitney test, it is reasonable to conclude that expensive bikes last longer.

CHAPTER 6

1. (a) H_0: Violent cartoons do not encourage antisocial behavior. In my opinion this is a very dubious null hypothesis, but some would regard it as appropriate. H_a: Violent cartoons encourage antisocial behavior. A one-tail test is appropriate.

(b) H_0: Barely plausible for me, others may find it quite believable. H_a: Very plausible for me.

(c) Cost of Type I error: Provides ammunition in favor of unnecessary restriction on civil liberties (censorship of violent cartoons). Cost of Type II error: Results in missing an opportunity to argue for effective intervention (censorship of violent cartoons) that will reduce antisocial behavior in the next generation.

(d) For me, the benchmark p value would be 1. Since I believe that H_a is more plausible than H_0, I am going to believe in H_a even if results can be readily explained by H_0. My belief that the costs associated with a Type I error are less important than the costs associated with a Type II error reinforces my decision to adopt a benchmark p value of 1. Others may believe that the traditional benchmark p value of 0.05 is appropriate.

(e) $p = 1/16 = 0.0625$. I am already convinced of H_a. Others may still be unconvinced, but some who were initially skeptical should be convinced by this.

2. Answers are subjective. I give those from my own world view. The following
   hypotheses are inappropriate: b, e, g, i, k, m, n, q, u, v, z through to ee, gg, kk, and
   mm. I believe a priori that these null hypotheses are probably false, some almost
   certainly so. All the other null hypotheses are appropriate in my opinion. The
   following hypotheses are in my opinion a priori highly believable and so I would
   require a very small p value to talk me out of them. In terms of the question, I
   would require a large value for x, as in 1 chance in x (for example, 1 chance in
   1,000,000 or $p = 0.000001$). Note that I express opinions from a skeptical point of
   view. Believers in astrology, for example, may have quite a different and valid
   viewpoint about including answers d and oo in the following list. Believers should
   go further and regard it as inappropriate to test hypotheses about astrology from
   the starting point that they don't work. However, believers may still consent to
   these null hypotheses to present evidence to convince skeptics. My list is a, d, x,
   ii, jj, oo, pp, rr, ss, and uu. Opinions regarding an appropriate p value may vary for
   many of the other null hypotheses.

3. Note that some would regard many of the H_0s listed in this answer as being suitable for a one-tail test to be inappropriate H_0s, as they would have a priori ideas
   that the H_0s were almost certainly false. If one did not believe testing was inappropriate, a one-tail test would be used for a, c, d, e, h, i, and k. A two-tail test
   would apply to the remainder.

4. Std error 0.2670; t_5 values corresponding to 5%, 2.5%, and 0.5% probabilities in a
   single tail are 2.015, 2.571, and 4.032, respectively. Hence, (a) 90% confidence
   interval (201.662, 202.738), (b) 95% confidence interval (201.513, 202.886), (c)
   99% confidence interval (201.123, 203.277), and (d) it seems more reasonable to
   believe that the bars are on average heavier than 200 unless there are strong a
   priori reasons to believe they are not. In any case, while it may be necessary to
   use 200.0 or some other single figure for the mean, we should recognize that there
   is zero probability that the true population mean *exactly* equals a single nomi-
   nated figure.

CHAPTER 7

1. Many examples are possible. For example, obesity and lack of exercise are likely
   to be associated because lack of exercise results in obesity: Food eaten is not
   needed to fuel muscle power and so is stored as fat. But obesity also makes it
   uncomfortable to perform exercise, and social factors may also make the same
   person less motivated to look after his or her health, both in terms of maintaining
   normal weight and in terms of keeping fit. Other examples include many circumstances where the expression "vicious cycle" is applicable.

2. No. There is overmatching.

3. The possibility that looking at football rather than music causes heart attacks seems
   farfetched. On the other hand, it may well be reasonable to believe that the two
   are associated as an example of a case of C causing both A and B. The social
   position and lifestyle of a football enthusiast is likely to be very different from
   that of a classical music enthusiast, and the former may tend to smoke more, eat
   more fatty food, or have more of other risk factors for heart attacks than the latter.

4. Knowing of no reason why there would be an association between cholesterol and violence, I would want more compelling evidence than the benchmark p value of 0.05. On the other hand, the brain is presumably affected by blood chemistry and so I would not want to blame evidence of an association on chance when it is exceedingly hard for chance to explain the results. I'll compromise and choose a p value of 0.01. Sign test p value (two-tail) gives 0.0026. Therefore, I now prefer to believe that low cholesterol is associated with a violent death.

Learning that the test for an association with violence was just one of many statistical tests that were conducted causes me to revoke my original opinion. If one looks for long coincidences for long enough, one is bound to find them. In other words, finding one long coincidence during a search of many possible coincidences is no long coincidence. Given that the criteria for deciding that an effect is "for real" is the occurrence of a genuine long coincidence, it is appropriate to now not believe the effect is "for real."

Note, the wording of the question taken together with the material in Chapter 4 would suggest the use of the sign test. However, the sign test is not quite appropriate here. The null hypothesis used in the sign test assumes that the probability of above or below median cholesterol is 0.5000 for each person counted as a violent death. This is true for the first person counted here. However, if the last person to be counted has a below median cholesterol, they would have been chosen from a group of 1,981, 984 of whom had below median cholesterol (19 people out of the original 2,000 having already been eliminated). The probability of choosing this last below median cholesterol person is then 0.4967, not 0.5000. The more appropriate test is Fisher's exact test, mentioned in the previous question and in Chapter 3 and discussed further in Chapter 8. However, with an experimental group of 2,000, of whom only 20 are removed, there is little difference between the p value 0.00258 given by the sign test and the p value 0.00246 given by Fisher's exact test.

If 1,800 have died, our analysis is not changed but the interpretation is. Since high cholesterol is associated with premature death from heart disease and nearly everyone has died, it may well be that those who died from violence died that way because they had lived a long time and so had longer exposure to risk of violence than those with high cholesterol who were likely to have died prematurely. Low cholesterol then causes a long life, which causes increased risk of violent death, but cholesterol itself does not increase the risk of violence.

5. (a) One may detect rubella as a cause, but there is concern that false associations will be detected because mothers of retarded babies might be more introspective about their pregnancies and recall more incidences, whether they are relevant or not.

(b) The study should be able to detect that having many children is a risk factor for parenting a mentally retarded child. Note that it is logical to expect that the mothers of retarded children will have more children on average than their neighbors and the rest of the female population. This is true even if being one of many children is not a risk factor for retardation, simply because a mother of ten, say, has given herself ten chances of producing a retarded child whereas a mother of one has only given herself one chance.

(c) Age effects may be detected by this study design. However, there are at some complicating factors. As explained in (b), mothers of retarded children are more likely to be mothers of many children. It takes time to produce many children, so these mothers may be older. It is also possible that there are problems with the control group. These are the altruistic neighbors. There could be a tendency for the amount of altruism in the control group to be associated with their age, so perhaps the control group might differ in age from the "ideal" control group.

(d) and (e) There is overmatching. If one house is poor or near a smelter, the same is likely to apply to the neighbors.

CHAPTER 8

1. Fisher's exact test $p = 0.0961$ (two-tail), 0.0514 (one-tail). Since I wouldn't need much convincing that an imposing stature is an advantage in administering others, I would set my benchmark p value somewhat higher than the conventional 0.05 and use a one-tail test. I therefore would now be reasonably convinced that tall stature is a factor in becoming an administrator.

2. Fisher's exact test $p = 0.00858$ (two-tail), 0.00429 (one-tail). Since H_a is not implausible, it now seems reasonable to believe that absentee ownership is associated with erosion.

3. Mann–Whitney two-tail test $p = 0.00909$. Using half this value, the result of the one-tail test is appropriate, because a drug designed to strengthen may not work but would be most unlikely to weaken. An independent samples t test is inappropriate because the values seem unlikely to be even approximately normally distributed, but this test gives a similar result. There is the same result for men and women: The drug strengthens both. The combined result for men and women is entirely unconvincing, with a (two-tail) p value of 0.6033 (approximate method; $p = 0.6297$ by the exact method). This is an example of Simpson's paradox, but where the outcomes are numerical measures rather than categories. All men are stronger than women (in this example). The drug makes both men and women stronger, but women with the drug are still weaker than men without the drug, and most women get the drug and most men do not. The combined result reflects the fact that the advantage of the drug to both sexes is swamped by the tendency of the weaker sex to get the drug.

4. Chi-square test of association $p = 0.0114$. It seems plausible to think that people with different belief systems may have different priorities. On the other hand, one wouldn't want to make generalizations about this on flimsy evidence. The conventional 0.05 benchmark seems reasonable to me here. Therefore, after seeing this calculation I would believe that there is a real association here.

5. McNemar's test (calculated as a sign test with eight in "favor" and twenty-two "against" comparisons) $p = 0.0161$ (two-tail). Halving this p value to give the results of a one-tail test would be appropriate if one knew that there were biological reasons to suspect carpet in causing asthma, or if one was aware of a campaign to persuade the parents of asthmatic children to remove carpet from the child's bedroom. This is convincing evidence of an association between carpet and asthma. Fisher's exact test of 152 asthmatics with carpet, 48 asthmatics without carpet, 138 nonasthmatics with carpet, and 62 nonasthmatics without carpet gives $p = 0.1453$ (two-tail); χ^2 test on same figures gives $p = 0.117$. The signifi-

cance of carpet is partly submerged under the weight of the many cases where both children or neither child has carpet as part of the family lifestyle. Figures of 8 and 22 added to these much larger numbers will give results that can easily be explained away as chance fluctuations in large numbers.

6. Use the chi-square goodness of fit test with expected values of 100 for each of the six outcomes. The chi-square value is 2.68, and values at least this far away from 0 in a chi-square distribution with 5 degrees of freedom occur 0.75 of the time (i.e., $p = 0.75$). There is no convincing evidence that the die is unfair.

CHAPTER 9

1. ANOVA gives a p value of 0.0451. If we were to use the conventional benchmark p value of 0.05, we would say that there is evidence that marks in statistics are related to the day of birth. Of course, this is silly. If I were offered the choice of believing that there was a mysterious reason at work that linked day of birth to statistics performance or instead believing that a 4.5-percent chance had eventuated, I would certainly take the latter. In other words, my benchmark p value would be much lower than the traditional 0.05.

2. ANOVA gives a p value of 0.549. If we were to use the conventional benchmark p value of 0.05, we would say that there is no convincing evidence that the improved varieties have a yield any different from the traditional variety. However, if we look at the means of the three varieties, we see that the improved varieties have higher mean yields than the traditional variety. It is true that these additional yields could easily be accounted for by chance given the amount of variation here, but it may be unfair to blame chance for the result when we know that the improved varieties have been produced to give higher yields. It would certainly be untrue to conclude that the figures show no evidence of improved yields: They provide some evidence, but it is evidence that could be easily accounted for by chance. It should be noted that the farmer's misleading words "no evidence" here are used in similar situations by many who use statistics. It is noted that the amount of data is small and it seems likely that if the farmer was to obtain more data he would eventually obtain enough to give a convincing p value. However, there is no obligation on the farmer or others who present conclusions based on statistics to use enough data for all worthwhile differences to emerge.

3. ANOVA gives a p value of 0.164. The same argument applies here as in the answer to question 2, but even more strongly. This really is an inappropriate null hypothesis. It is general knowledge that locations further from the equator are colder on average. We should be convinced of this no matter how easy it is for chance to explain results on the assumption that all four locations have the same average temperature. However, there is a further point here. ANOVA is not the optimum method for analyzing these data. It doesn't use the information given about latitude. This information can be used if we think of the data in the form of the following list:

Latitude	19.5	19.5	27.5	27.5	34	34	37.75	37.75
Temperature	27	31	24	18	16	22	13	21

Regression can be applied and it gives a p value of 0.0165 for the hypothesis that there is no relationship between latitude and temperature. Although we should already have been convinced that there was a relationship between temperature and latitude, we can be convinced yet again by this analysis.

4. Regression $p = 0.0000664$. It still seems more likely that what we are seeing in these figures is the outcome of a 6 in 100,000 chance rather than evidence that a mysterious force stemming from faraway Brisbane's rainfall is having influence on American book importers. In other words, in these circumstances my benchmark p value is way below the conventional 0.05 and would be considerably less than 6 in 100,000.

5. Regression $p = 0.114$. This tells us that the variation in house prices in 2000 could "quite easily" be explained by chance without any relationship to the value of the houses ten years earlier. However, I would be surprised if the value of the houses previously had no bearing on their value now. I would either regard this as an inappropriate null hypothesis or else allow myself to be talked out of this null hypothesis very easily. Noting that the β is positive, indicating that as the original price tends to increase the recent price tends to increase, I would think it far more believable to think that this was a real relationship rather than the product of a chance that happens 11 out of 100 times.

6. 1 time in 6! $= 720$ times. Yes, convinced by Spearman's rho of -1 with $p = \frac{1}{720} \approx 0.0014$.

7. The regression effect. The very wettest places one day might often be wet, but are not that likely to be the wettest places another day.

8. Sign test $p = \frac{1}{512} \approx 0.002$. The regression effect again. The places with the highest number of rabbits one year may well tend to have a rather high number of rabbits most of the time but will not usually be the places with the highest numbers every time. They were the highest on the first occasion presumably not only because they provided good habitat for rabbits, but also because chance factors had particularly favored the rabbits at those places in that year. These chance factors will normally not boost rabbit numbers to the highest levels in other years.

There is another issue that applies to questions (7) and (8). The data will not generally be independent, as is assumed by our statistical tests. Rabbit numbers everywhere at the time of the second measurement might be down because the second measurement might be taken during a drier year than the first.

CHAPTER 10

1. (26.739, 27.261) using the normal distribution as an approximation to the t_{99} distribution; (26.735, 27.265) using t_{99} distribution itself.

2. (a) (68.579, 70.841). (b) The confidence interval is of zero width about the point 69.71, since we know the average of the marks of this class with certainty.

3. (a) 16,587.

(b) It is true that when we are about to take our sample we will have at least a 99-percent chance of obtaining a value within 1 percent of the true value ("at least" applies because we use the largest possible value of the product θφ in our theory). It is not true that a particular value obtained necessarily has a 99-percent chance of being within 1 percent of the true value. After obtaining a particular outcome we can't reverse the probability statement made prior to the survey to give a probability statement about the location of the true mean. Instead, we will have a confidence interval (here it is [39%, 41%]). As explained in the section on confidence intervals, confusing confidence intervals with probability statements is like confusing the statement, "Where there is smoke there is a 99-percent chance of fire," with the statement, "Where there is fire there is a 99-percent chance of smoke." The range from 39 percent to 41 percent consists of those numbers for the true proportion that are compatible with our findings in terms of a p value of 0.01. However, we may feel that sheer bad luck has given us a misleading result and we need to take this into account. If we feel our survey has underestimated or overestimated the true proportion, then there is not a 99-percent chance that our range contains the true proportion. Only in the absence of a prior opinion about the true value is it reasonable to equate confidence intervals with probability intervals.

4. (a) Fisher's exact test or chi-square test of association

    (b) ANOVA

    (c) Independent samples t test

    (d) Mann–Whitney test

    (e) Sign test

    (f) Paired samples t test

    (g) Wilcoxon signed rank test

    (h) Fisher's exact test or chi-square test of association

    (i) ANOVA

    (j) Chi-square test of association

    (k) Regression

Annotated Bibliography

There are many books that can be read by people seeking further information in this field. There are several main areas: Books that give more general mathematical background, books that give more of the theory of probability and statistics, and books on applications of statistics. What follows are a few books I have found helpful in these areas.

C. V. Durell. *Advanced Algebra*. Vol. 1. London: G. Bell and Sons (first published 1932 and reprinted numerous times). *My book started with a promise that there would be no mathematics beyond the tenth grade level. I believe I have kept that promise, though there are places in the book where I indicate that a deeper knowledge of mathematics would enable a more complete understanding of the details of the statistical techniques. There is a great deal that would have to be covered for a complete mathematical understanding of all the steps used in deriving all the statistical techniques covered in this book. However, as a start, some twelfth grade mathematics would give readers more insight in parts of this book where I have referred to logs and exponentials and the number e. There are many texts on twelfth grade mathematics. Different readers have different learning styles and so there is no such thing as an ideal book for everybody. Personally, I have found this little old book by Durell to be very useful. Its clear but very concise explanations appeal to me.*

Stephen Jay Gould. *The Mismeasure of Man*. London: Penguin Books, 1996. *This is a general-interest book rather than a textbook. It deals with one major area in which statistics has been misapplied. This is the ranking of humans and groups of humans according to mental ability. It contains numerous examples where statistics are flawed at various stages, ranging from the selection of a sample to the interpretation of the results. However, these flawed statistics deeply influenced much of the social policy and education system of the Western world over the last century. The book contains an interesting discussion that hints at* the richness as well as the potential for misuse of more advanced statistical theory.

J. A. Rice. *Mathematical Statistics and Data Analysis*. 2d ed. Belmont, Calif.: Duxbury Press, 1995. *This book covers more briefly the material on probability relevant to statistics that is covered by S. Ross. It then goes on to give an almost complete mathematical understanding of all the steps used in deriving all the statistical techniques covered here. However, the book assumes knowledge of calculus and the theory of matrices at the first- or second-year university level.*

Sheldon Ross. *A First Course in Probability Theory*. 3d ed. New York: Macmillan, 1988. *This reasonably slim book gives a more complete understanding of probability theory and random variables. Some knowledge of calculus at the twelfth grade level is necessary for understanding some sections in the latter parts of the book. In one section more advanced knowledge of calculus is required.*

Index

ABOUT THE AUTHOR

David Kault is a medical practitioner and an adjunct lecturer in mathematics at James Cook University, Queensland, Australia. He has taught a number of introductory statistics courses, both general and applied to such areas as environmental science and medicine.